SpringerBriefs in Sociology

For further volumes:
http://www.springer.com/series/10410

SpringerBriefs in Sociology

Aaron A. Toscano

Marconi's Wireless and the Rhetoric of a New Technology

 Springer

Assoc. Prof. Aaron A. Toscano
Department of English
University of North Carolina
9201 University City Blvd.
Charlotte
NC 28223-0001
USA

ISSN 2212-6368 e-ISSN 2212-6376
ISBN 978-94-007-3976-5 e-ISBN 978-94-007-3977-2
DOI 10.1007/978-94-007-3977-2
Springer Dordrecht Heidelberg New York London

Library of Congress Control Number: 2012931951

Printed on acid-free paper

Springer is part of Springer Science+Business Media (www.springer.com)

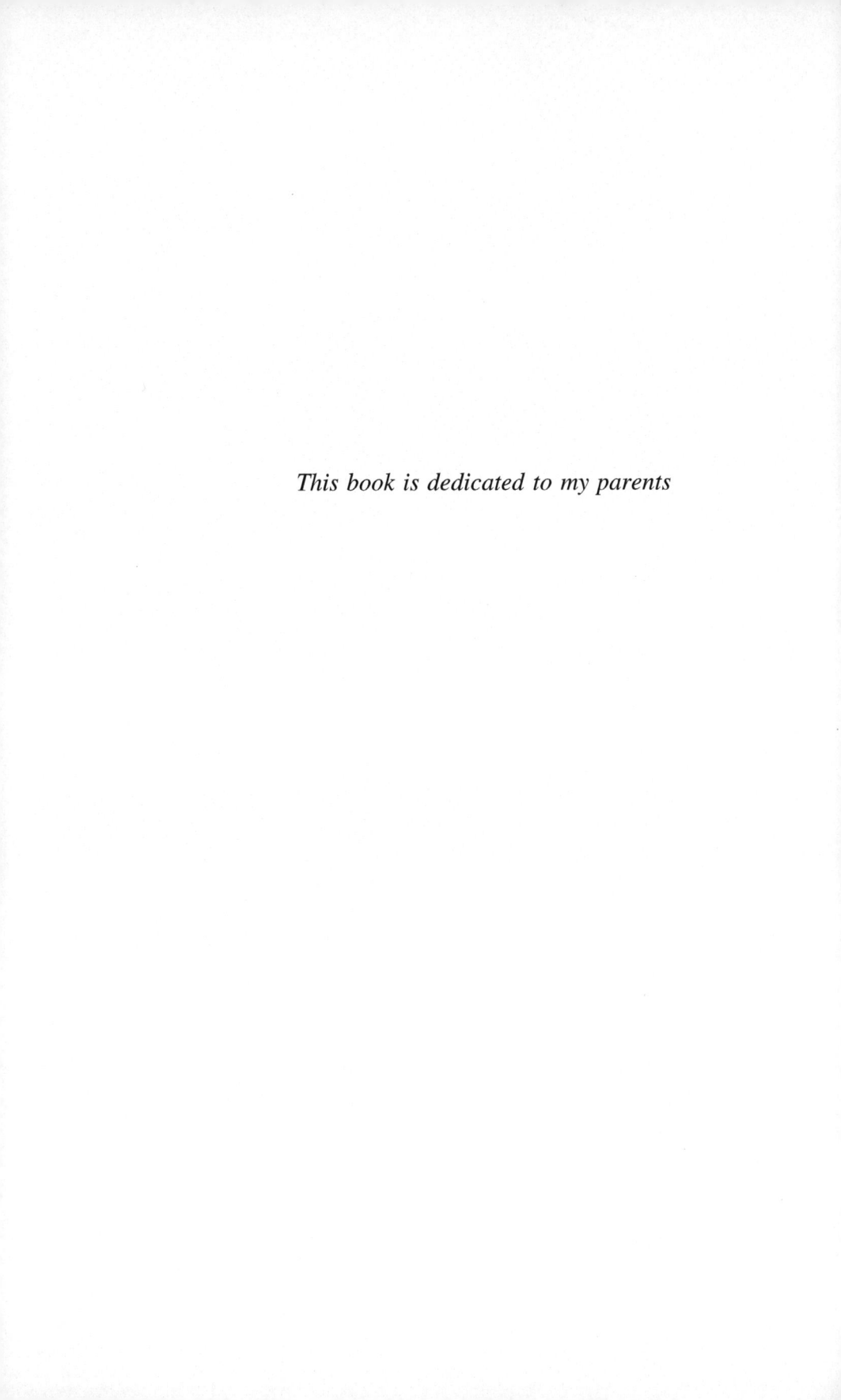

This book is dedicated to my parents

Acknowledgments

Just as no technology is created in a vacuum, no book is written without the influence of others. I would like to thank the following individuals for their support: Debra Journet, Mary Rosner, Aaron Jaffe, Dennis Hall, Avery Kolers, Alan Golding, Suzette Henke, Mark West, Malin Pereira, Jeffrey Leak, and my friends and colleagues at both the University of Louisville and the University of North Carolina at Charlotte. Additionally, I have to thank my Springer representatives, Miranda Dijksman, Esther Otten, Myriam Poort, and Hendrikje Tuerlings, for seeing this book through to publication. I might be missing some others and, for that oversight, I do apologize.

I also feel it is important to acknowledge friends and family (you know who you are) who helped in tangible and intangible ways. Because I plan to devote my life to teaching and research, I have to thank the many wonderful teachers I've had the privilege to study under. We are the culmination of our experiences, and the people with whom we interact, whether or not we always notice it, play a role in constructing our identities.

Contents

Introduction

As Sir Oliver Lodge has stated, [using wireless signals to cross the Atlantic] was an epoch in history. I now felt for the first time absolutely certain that the day would come when mankind would be able to send messages without wires not across the Atlantic but between the farthermost ends of the earth.

(Guglielmo Marconi 1901a/1999, p. 32)

Guglielmo Marconi made international headlines when he successfully sent and received wireless signals across the English Channel on March 27, 1899. This significant event marked the first international wireless communication. Many important English and French journalists observed the event and promoted Marconi as an international celebrity much like Thomas A. Edison and Alexander Graham Bell. Less than three years after crossing the English Channel, Marconi solidified his celebrity status by transmitting wireless signals across the Atlantic Ocean on December 12, 1901. Newspapers around the world reported what had happened between Poldhu in Cornwall, England and Signal Hill in St. John's, Newfoundland—the Atlantic Ocean had been crossed without using wires. Marconi was awarded the Nobel Prize for Physics in 1909 for his work in wireless communication.

Marconi was a technologist with a shrewd business sense and enough electrical engineering skill to represent the wireless as an important advancement in science and technology in the early twentieth century. Marconi used towers, balloons, and other related wireless components he compiled or improved in order to physically construct his invention. Additionally, Marconi represented the wireless through charts and scientific and engineering discourse to the electrical engineering community. Through this discourse, Marconi proposed the wireless as a *possible* new technology. His language appealed to the audience's values—rigorous scientific practice, need for detail, potential impact—in order to show that the wireless was a viable technology. This language is a form of technical communication. Contrary to the contemporary bias that technical communication is simply documenting procedures or writing engineering reports, Marconi and others communicated the idea of his wireless using rhetoric mediated by modernist ideology. He and others described the wireless, thus, acclimating audiences to its meaning, by describing the wireless's attributes in relation to the time period's values. Audiences learned about the science of wireless transmissions *and* its cultural message through technical communication. All discourse surrounding technology is technical communication, and this book argues for broadening the scope of technical communication pedagogy and scholarship to include attention to the types of discourse similar to the discourse that rhetorically constructed the wireless. Currently, consumer technologies are heralded as must haves on TV, the Internet, and the magazine rack. Ignoring this discourse limits the field's chances to examine all types of discourse that introduce users to technology.

Marconi's invention demonstrates that technological advances are not merely created out of thin air. Technologies are inventions and innovations created by physical, rhetorical, and mechanical actions. Also, a technology is both a product and distinguishing characteristic of any historical context: Social conditions help create new technologies, and society itself is shaped by these contexts (Feenburg 1999; Giddens 1984; Lomask 1991; Nye 1994). To understand how technologies become realities, we must examine how they are promoted, negotiated, and constructed. While technological blueprints show how a technology is schematically represented, they do little explicitly to reference the historical and cultural context in which the technology was created. Also, understanding the social situations present during a technology's creation allows us to discover the relations among science, technology, and society.

Scholars critically analyze the rhetoric, philosophy, history, and sociology of technology and science in order to demonstrate the social construction of scientific knowledge and technical artifacts. This work falls under the broad category of Science and Technology Studies (STS). STS seeks to understand technology and science from a cultural perspective by investigating both discursive and physical situations that mediate the teams of engineers and scientists making "discoveries." The main contribution to the field or, more accurately, *fields* has been recognition that technologies and sciences do not come from laboratories isolated from society; instead, for technologies or sciences to be realized, they must fit or be made to fit within established cultural values and practices. STS attempts to demonstrate how ideological and individual values negotiate which discoveries become realities and what social values "demanded" these technologies.

One process used to understand how technology and science reflects (often dominant) social values is to examine discourse surrounding an invention or scientific discovery. Technology and science are constructed based on social values and through social interaction. Rhetoric is important for this endeavor because new technologies and sciences are presented to various audiences not just as physical apparatus but also as discourse—the main social interaction among producers, sponsors, regulators, and users. Regardless of specific focus, rhetorical analyses of technology or science aim to uncover the methods and tactics used to construct the reality or image of a new discovery. Because social values appear within technological contexts, we can "read" technologies similarly to how we read cultural works—art, literature, film, etc. The social context affixes meaning to a technology, and a technology may also become a mark of the time period, civilization, or group. For instance, the "Bronze Age" defines a time period, stretching across cultures, where the dominant metallurgy processes created bronze instruments. Bronze Age peoples created non-bronze technologies, but historians characterize the era based on its peoples' use of a particularly important technology. Significant technologies can define a time period. Thomas Edison's invention of the electric light bulb and Guglielmo Marconi's invention of the wireless radio stand as two of the modern era's most significant technological advances. As Bazerman (1999) argued, advances in technology are framed and defined by language: "[t]he invention is legally not a physical entity. It is a symbolic representation—a text representing an idea" (p. 91). Bazerman does not just mean patent documents are the texts symbolically representing an invention; any discourse—for instance, Marconi's presentations—is a "symbolic representation" that also acts as an "ideologically saturated" medium. This discourse is technical communication that rhetorically represents an invention's meaning to society.

Beyond the ideologically saturated words an inventor/author uses to describe any technology is the technology itself, which is often simply an idea swirling in a dialectical frenzy of social, rhetorical, and physical negotiations. Texts and speech acts enable audiences to interpret, reinterpret, and question the need for the technologies that enter their culture. People learn about technologies not only by actually working with them but by interacting with their representations. Bazerman (1998) identified that "technology has always been fundamentally designed to meet human ends," but adhering to those ends requires "technology, as a human-made object," to be "articulated in language and at the very heart of rhetoric" (p. 383). The wireless was a technology conceived by an audience through discourse before users encountered the physical "black box." A realized technology or science is a "black box"—a technology about which debate has ceased—which is Bruno Latour's (1987) metaphor, developed from the term cyberneticians use to signal whenever a piece of machinery or set of commands is too complex, that describes scientists' and engineers' approaches to "established" facts (pp. 2–3). Any science that is a black box is *assumed* to be unalterable. Scientists and engineers risk devoting too many laboratory resources if they attempt to debunk or simply reexamine black boxes. New science and technology can be constructed based on the premises established by a black box (e.g., Watson and Crick's "discovery" of the double-helix structure of DNA).

This rhetorical analysis examines how Guglielmo Marconi's wireless became realized as a product of modernity. The wireless eventually became the black-boxed *radio*, a fully realized technology, but, before it did, Marconi's technology was presented to technical and scientific audiences, *re*-presented in favorable popular press articles, and reinscribed into F. T. Marinetti's early Futurist aesthetics along with other "progressive" technologies.[1] In this book, I limit my examination to favorable representations of the wireless in order to concentrate on how authors attempt to excite their readers or convince them that Marconi's wireless was an important reality. These sources suggest much about how the wireless stood as a marker of civilization. Although there were certainly negative portrayals of the wireless, as there have been for many technologies, my analysis concentrates on how stakeholders and relevant social groups use language to build meaning into a successful technology, one that reflected the assumptions of an era.

My research into Marconi's contemporary discourse and rhetoric provides examples of how "progress" and related attributes accompanied most descriptions of the wireless. In fact, having the label "progressive" in industrial cultures means that a technology, individual, or nation is advanced economically, evolutionarily, and even militarily. Inventors, journalists, marketers, and users affix progress labels on new products. The discursive representations of technologies create the sense that the technologies physically exist. Often times, though, the representations are merely projections of a technology's possibilities—the technology could seem real based on a rhetorical act. Journalists report on these images and, subsequently, audiences believe the technology exists before a viable product makes it to the marketplace. Investors, consumers, or readers must believe in a technology's viability before it can be said to exist. Marconi's wireless existed as an idea prior to his major commercial success. The wireless began as a compilation of gadgets from other inventors working on similar wireless systems. Marconi's genius is assembling these components into a viable technology. Before he had a viable product, though, he would position the wireless as a forthcoming monumental technology—one marking itself as important for Western Civilization. Much of Marconi's positioning of the wireless could be considered simply shrewd marketing, but the rhetoric of the wireless holds cultural values as well. In other words, the time period influenced the discourse surrounding the wireless.

This book's focus is on the ideology of the time period and not Marconi's biography. Most research on Marconi has been in the form of biographical sketches of his life (Corazza 1998; Garrat 1994; Hong 2001; Marconi 1982, 2001; Tarrant 2001) or historical accounts of the wireless as the precursor to radios and related communication technologies (Bucci et al. 2003; Crowther 1954; Jensen 2000; Kraeuter 1990). The historical and biographical scholarship on Marconi's wireless describes the wireless in a detailed time-line fashion and portrays Marconi as a genius inventor. While such studies are important for narrating the wireless's physical construction and Marconi's life (I rely on both in my study), they perpetuate the "lone inventor" myth. While I acknowledge the fallacy, I choose to concentrate on the positive representations of his particular commercial invention as a way to uncover the time period's attitudes and values. He was not the sole inventor, but he was, arguably, the most important historically. Certainly, Marconi held a high celebrity status in the early twentieth century. After crossing the Atlantic Ocean, as one article claims (Baker 1902), "[t]he people of the 'ancient colony' of Newfoundland ...*crowned* [Marconi] with every honor in their power," and that "it seemed as if every fisher and farmer in that wild country had heard of him ... at twenty-three he was famous the world over" (p. 4, emphasis added),[2] despite the fact that his fame was probably concentrated in Europe, America, and the colonies. McClure (1902) portrays Marconi's elevated status by noticing that

"Marconi's party occupied four staterooms on the upper deck" during wireless experiments aboard the *Philadelphia* (p. 526). Marconi and his "party" carried out the experiments in one of the rooms, but occupied the other "upper-deck state-rooms" as would any high-class celebrity.

The existing research credits Marconi with bringing a monumental technology to life, and this work is useful for understanding the historical significance of Marconi's wireless. Research has been done on the effects of radio on particular communities after the wireless became entrenched in national (and world) infra-structure (Patnode 2003; Rutland 1994; Squier 2003), but no rhetorical analysis has been done on the wireless, and only recently has anyone examined the rela-tionship between the wireless and modernist aesthetics (c.f. Campbell 2006). Also, no scholarship considers what cultural work the wireless did based on contem-porary descriptions. This book focuses on the rhetoric of the wireless and adher-ence to cultural values and leaves issues of Marconi's personal life, business savvy, and agency to the rich biographies mentioned above.

The rhetorical analysis this book engages is an important cultural studies approach for the field of technical communication (or technical writing). Technical communication is a culturally mediated discipline that communicates ideas about technology. One goal of this book is to demonstrate to the field of technical communication how to incorporate rhetorical analyses of technology into its research and pedagogy. The analysis I carry out on the language surrounding the wireless is really an analysis of technical communication. Chapter 1 broadens the definition of technical communication to include all discourse surrounding tech-nology. Many assume that technical communication deals solely with such doc-uments as specifications, manuals, technical reports, etc., but technical information as well as information on technologies in general can be found in a variety of texts. Both new and old media communicate technical information and information about technologies. That discourse is not usually considered when defining tech-nical communication, but, because it describes technologies, it is a type of tech-nical communication. No technology or communication endeavor is created in a vacuum. Communication about technology necessarily draws on community-defined assumptions influencing the format as well as the essence of the discourse surrounding technology and, therefore, has a rhetoric. Technologies and commu-nication about technologies already have implicit values because both are medi-ated by ideology. The subtext of any communication (or speech act) is rhetorical, adhering to the attitudes and practices of senders and receivers. To demonstrate how culture mediates technical communication, Chap. 1 briefly traces the schol-arship on the history of technical communication from the English Renaissance to today, describing the rhetoric of technical communication across the centuries.

In order to have a sound theoretical framework from which to argue the wireless's social, rhetorical, and aesthetic values, Chap. 2 reviews important STS theories and case studies that show technology and science to be products of social interaction. I first discuss theorists such as Charles Bazerman, Bruno Latour, and Wiebe E. Bijker who argue that groups affix values to new technological advancements. Next, I explain how rhetoric contributes to the non-physical ways in which technologies are constructed for audiences. After all, a technology's rhetoric helps establish a technology as a product congruent to social values, attitudes, and practices. Because Marconi's wireless fit the early twentieth century's progressive ideology, I discuss how new technological advancements in the twentieth century conformed to modernist ideals of industrialized nations. These arguments help me demonstrate the importance of rhetorical analyses of technology in general and the rhetoric behind Guglielmo Marconi's wireless specifically.

Chapter 3 analyzes the topoi Marconi used when presenting the wireless to a technical audience. Marconi's rhetoric promotes the wireless's viability and necessity for members of the Royal Institution and the Royal Society of Arts, London. Before the wireless became the radio, Marconi prophesized its usefulness, linking it to cultural values. The wireless embodies certain industrial traits—speed, efficiency, profitability, evolution/advancement—and these technical presentations employ such images. Even in the most technical forum, Marconi promotes cultural pride and includes the wireless's social and economic characteristics alongside discussions of its technical viability. At the time of Marconi's presentations, the wireless was not yet a major commercial product; instead, it was an idea for linking ships at sea, ships to land stations, and even nations with one another. In order to have a successful product, the wireless had to be portrayed as possible and *progressive*. The need for instant communication had been around for nearly 40 years as evidenced by telegraph and telephone technology. The wireless did not replace those technologies, but it did expand the reach of communication and, therefore, mass communication. As an important mouthpiece for the wireless, Marconi offered his audiences not the physical product but the idea of a potential product to mark human advancement and bolster economic progress.

Chapter 4 demonstrates how Marconi's wireless was re-presented to audiences through favorable popular press articles in American periodicals. Many journalists reconstructed the wireless rhetorically by emphasizing the technology's industrial value and profitable potential. The rhetoric the authors employed closely follows the topoi Marconi used in his presentations to technical and scientific audiences. The articles promote the wireless's efficiency, profitability, and usefulness by using progress markers. The popular press also portrayed the wireless as a technology marking human evolution. Although the texts I use do not represent all early twentieth-century audiences, the descriptions are at least suggestive of positive discourse concerning the wireless. After all, the wireless became a successful black box—the radio—so the re-presentations suggest how positive accounts of the wireless fit the cultural values and attitudes concerning technology in the early twentieth century. Progressive ideology of the early twentieth century is entwined with these favorable representations. Although these positive popular press accounts were not the only types of descriptions about the wireless, they illustrate what Marconi's contemporaries had to say about an ultimately successful technology, an invention heralded as a monumental invention from the turn of the last century.

Chapter 5 demonstrates F. T. Marinetti's glorification of early twentieth-century technologies such as the wireless. Marinetti saw the wireless as an important tool and marker of human advancement. More than any other modernist artist, Marinetti shows the "virtues" of new technologies. Marinetti wanted his audiences to embrace the values of progress as represented through speed, efficiency, evolution, and ahistoricity. Machines appear as muses to Marinetti who fantasizes about becoming one. Such a transformation would make Marinetti the most efficient being possible: By casting aside his human qualities and "softness," he would be free. Marinetti's art makes use of *parole in libertà*—words in freedom—to accentuate the aesthetic goals of minimalism and telegraphic prose. He advocates an aesthetic based on reducing expressions to the fewest number of words. Within this minimalist aesthetic, Marinetti promotes the speed and efficiency of wireless communication, using words not attached to traditional syntax. Similarly, Marinetti claims the wireless is an inspiration for such a telegraphic style. Marinetti's art reconstructs the cultural values of industrialization by promoting textual efficiency. Even in Marinetti's art there is something technical about his desire for words to capture efficiently the essence of the technology, idea, or, most importantly, action. Also, Marinetti glorifies technologies for their speed and war potential. Along with the wireless, Marinetti groups together the following technologies he felt defined modernity: tanks, airplanes, machine guns, and automobiles. Unlike other high modernist authors, such as D. H. Lawrence and Virginia Woolf, Marinetti values the destructive nature of new modern technologies. His manifestos exaggerate tropes of progress, advocating a love of mechanization.

The conclusion asks readers to think further about the rhetoric of contemporary technologies and how the various rhetorics help establish technologies as artifacts in accordance with cultural values. Even though an inventor or team of inventors may actually create a technology, societal forces fuel the technology's development. Technological descriptions must also follow the discourse community's conventions of rhetoric: Writers' descriptions of technologies form and are formed by the community's values and practices (Bazerman 1999; Journet 1993; Latour 1996; Longo 2000). For example, the wireless did not stand alone as a tool with a purely functional role; instead, the wireless's texts show that descriptions demonstrate how the technology fit into the culture. The wireless changed mass communication for the early twentieth century (Attwood and Ryecart 1997; Jensen 2000; Tarrant 2001), but first the wireless had to "fit" into the historical context of a technologically saturated world. We can locate society's values by examining how others praised the technology through the various discourses on the wireless.

New tools and techniques excited some individuals (i.e., Marinetti, Taylor, and the popular press writers I examine in this book) as did the World's Fairs or other major cultural events celebrating technology. These tools can be thought of as extensions or prostheses for human labor and evolution. Marinetti's (1913) "man multiplied by the machine" (p. 97) reflects the cultural condition of modernity that views scientific and technological progress as human evolution: Humans evolve by figuratively plugging into the new electric forces technology harnesses. For example, the wireless advanced human capabilities of speech by allowing an individual to communicate with others around the world instantaneously. Of course, at the time, communications were sent through Morse code, which constrained one's prose.

Marinetti saw such constraints as liberating words from syntactical "strings" by allowing one word to communicate an idea that took several words before. Such telegraphic minimalism was not unique to Marinetti; many artists and artistic movements experimented with a style that reflected industrial culture's drive for greater efficiency.

And artists were not the only group experimenting with "efficiency." The culture itself appears to value efficiency, which we read in the texts the time period created. Marinetti's texts share Frederick W. Taylor's (1911/1967) efficiency manifesto's goal for promoting increased production. Besides increased output, efficient technologies allowed life to speed up. Marinetti and others did not create the demand for efficiency by industrial societies of the early twentieth century; instead, they tap into a cultural imperative and, in Marinetti's case especially, amplify this already existing ideological tenet. Marinetti wanted speech streamlined as much as possible, and he saw the wireless as a practical tool for such a goal. Whereas Marconi and the popular press promote the wireless specifically as an important efficient technology, and Taylor promotes scientific management more broadly, Marinetti violently advocates efficiency as an artistic and even spiritual goal. The wireless and all technologies were tools for business, and industry, but Marinetti also believed they were tools for progressing away from a useless past. His work shows that technologies of the early twentieth century represented attitudes and values of a society, an industrial society. Because Marinetti is a product of modernity, as is the wireless, his work resides in the context of industrialization and mass culture. Within such a context, Marinetti's art conforms to other contemporary pro-technological discourses with tropes of progress as Marconi and the popular press employed. In short, his work reflects the culture's positive (albeit exaggerated) fascination that helps technologies become realized. The rhetoric of technology or, specifically, the rhetoric of the wireless presents this new technology as a tool for today *and* tomorrow. It is discourse more than the physical construction that identifies the artifact as existing.

This book does not attempt to show the physical construction of the wireless but to demonstrate how rhetoric created an image of the wireless that related to social, economic, scientific, and literary influences. Furthermore, this book is not a biography of Marconi or a close reading of his business acumen, which are both separate projects/studies outside the scope of this particular research agenda. Although Marconi's life and business vision are important factors in the wireless's invention and acceptance, there are other stories to tell about the wireless's construction. My focus on the rhetorical construction demonstrates how audiences can *read* early twentieth-century ideology in the discourse that promoted Marconi's wireless as a product of modernity based on a century of science. Technologies do not simply spring from the earth and change society; instead, human societies create an environment that allows certain technologies to be developed. What follows is one story of the wireless, a monumental cultural artifact of the twentieth century that still resonates today.

References

Attwood, D., & Ryecart, G. (1997). *The radio: An appreciation.* San Diego: Laurel Glen.

Baker, R. S. (1902). Marconi's achievement: Telegraphing across the ocean without wires. *McClure's Magazine, 18*(4), 4–12.

Bazerman, C. (1998). The production of technology and the production of human meaning. *Journal of Business and Technical Communication, 12*(3), 381–387.

Bazerman, C. (1999). *The languages of Edison's light.* Cambridge: MIT Press.

Bucci, O. M., Pelosi, G., & Selleri, S. (2003). The work of Marconi in microwave communications. *IEEE Antennas and Propagation Magazine, 45*(5), 46–53.

Bunch, B. (2004). *The history of science and technology: A browser's guide to the great discoveries, inventions, and the people who made them, from the dawn of time to today.* Boston: Houghton Mifflin.

Campbell, T. (2006). *Wireless writing in the age of Marconi (electronic media tions).* Minneapolis: University of Minnesota Press.

Corazza, G. C. (1998). Guglielmo Marconi—Marconi's history. *Proceedings of the IEEE, 86*(7), 1307–1311.

Crowther, J. G. (1954). *Six great inventors: Watt, Stephenson, Edison, Marconi, Wright, Brothers, Whittle.* London: Hamilton Press.

Feenberg, A. (1999). *Questioning technology.* London: Routledge.

Garratt, G. R. M. (2006). *The early history of radio: From Faraday to Marconi (IEE History of Technology, No 20).* Herts, United Kingdom: The Institution of Electrical Engineers (Original work published 1994).

Giddens, A. (1984). *The constitution of society: Outline of a theory of structur ation.* Berkeley: University of California Press.

Hong, S. (2001). *Wireless: From Marconi's black box to the audion.* Cambridge: MIT Press.

Jensen, P. (2000). *From the wireless to the web: The evolution of telecommuni cations, 1901–2001.* Sydney: University of New South Wales Press.

Journet, D. (1993). Biological explanation, political ideology, and 'blurred genres': A bakhtinian reading of the science essays of J. B. S. haldane. *Technical Communication Quarterly, 2*(2), 185–204.

Kraeuter, D. W. (1990). The U.S patents of Armstrong, Conrad, De Forest, Du Mont, Farnsworth, Fessenden, Fleming, Kent, Marconi, and Zworykin. *AWA Review, 5,* 143–191.

Latour, B. (1987). *Science in action.* Cambridge: Harvard University Press.

Latour, B. (1996). *Aramis, or the love of technology* (C. Porter, Trans.). Cambridge: Harvard University Press.

Longo, B. (2000). *Spurious coin: A history of science, management, and technical writing.* Albany: State University of New York Press.

Lomask, M. (1991). *Invention and technology.* New York: Scribner's.

Marconi, G. (1901a/1999). Messages without wires. In R. Rhodes (Ed.), *Visions of technology: A century of vital debate about machines, systems and the human world* (p. 32). New York, NY: Touchstone.

Marconi, D. (1982). *My father, Marconi* (2nd ed.). Ottawa: Balmuir.

Marconi, M. C. (2001). Marconi, my beloved (2nd ed.). In: E. Marconi (Ed.), Boston, MA: Dante University of America Press.

Marinetti, F. T. (1913/1973). Destruction of syntax—[Wireless imagination]—Words-in-freedom. In U. Apollonio (Ed.), *Futurist manifestos* (R. W. Flint, Trans) (pp. 95–106). Boston, MA: MFA Publications.

McClure, H. H. (1902). Messages to mid-ocean: Marconi's own story of his latest triumph. *McClures's Magazine, 18*(6), 525–527.

Nye, D. E. (1994). *American technological sublime.* Cambridge: MIT Press.

Patnode, R. (2003). What these people need is radio: New technology, the press, and otherness in 1920s America. *Technology and culture, 44*(2), 285–305.

Rutland, D. (1994). *Behind the front panel: The design and development of 1920s radios.* Philomath: Wren Publishers.

Squier, S. M. (2003). *Communities of the air.* Durham: Duke University Press.

Tarrant, D. R. (2001). *Marconi's miracle: The wireless bridging of the Atlantic.* St. John's, Newfoundland: Flanker Press.

Taylor, F. W. (1967). *The principles of scientific management.* New York: Norton (Original work published in 1911).

Chapter 1
The Rhetoric of Technical Communication

Maxims make one great contribution to speeches because of the uncultivated mind of the audience; for people are pleased if someone in a general observation hits upon opinions that they themselves have about a particular instance....A maxim, as has been said, is an assertion of a generality, and people enjoy things said in general terms that they happen to assume ahead of time in a partial way.

(Aristotle, trans. 1991, 2.21.15)

Technology rich societies have a need to communicate technical information. Industrial and post-Industrial societies rely heavily on technologies. Technologies range from the clothes on our backs to the Large Hadron Collider, but many citizens envision computers and communication technologies when they hear "technology." Technology (among a few other attributes) separates humans from other species. Humans are tool users and have passed down this knowledge using a variety of communication formats from oral to textual to multimodal. Although we cannot say, without qualification, that technology universally gets "better" because such a value-laden statement may refer to features held higher in regard for some than others, technologies come to exist in order to fulfill a need. Users may feel a new technology is better than an old one because the new technology reduces the time it takes to complete a task. Therefore, *better* means quicker or more efficient. In order to accomplish tasks, humans have created technologies—material and intellectual—to fulfill social demands. We can say with certain conviction that Industrial and post-Industrial societies expect technological solutions to fix or enhance human life. And being a member of technological society inevitably means one will not only use technologies but will both receive information about and communicate with technology.

The discourse surrounding technology—whether it be in a manual, peer-reviewed article, popular forum, or a variety of other media—falls under technical communication. Although technical communication has an ancient tradition, communicating about technology in the Information Age is a practice members of hi-tech societies cannot avoid doing and encountering. We formally and informally communicate ideas about and with technology: Whether we are advocating (or defending) using one mobile communication device over another or debating which energy sources the nation should pursue, we are engaging in communicating technical information. This communication does not always accurately represent the technologies, but we engage nonetheless because we have opinions about technology formed by our experiences and culture. In fact, even in the most formal

A. A. Toscano, *Marconi's Wireless and the Rhetoric of a New Technology*, SpringerBriefs in Sociology, DOI: 10.1007/978-94-007-3977-2_1, © The Author(s) 2012

technical situation, the communication is socially constructed and rhetorical: No technology or communication endeavor is created in a vacuum. Communication about technology necessarily draws on community-defined assumptions influencing the format as well as the essence of the discourse surrounding technology. Technologies and, therefore, communication about technologies already have implicit values because both are mediated by ideology. Every discourse community's ideology or values mediates their communication and epistemology; therefore, the subtext of any communication (or speech act) is rhetorical, adhering to the attitudes and practices of senders and receivers.

Although technical communication is often seen as non-rhetorical and didactic, it is still another ideologically mediated practice. This chapter asks readers to redefine technical communication as *all discourse surrounding technology* in order to prepare readers to identify the discourse surrounding the wireless as ideologically mediated in both technical and not-so-technical forums. Technical communication historically has been a vehicle for knowledge dissemination of technology- and science-related subjects, and this chapter briefly traces this history in the available technical communication scholarship. One goal for this chapter is to have readers (especially technical communication students and teachers) reflect on how contemporary technical communication is taught. Surveying the history of technical communication instruction is beyond the scope of this book. Instead, this chapter focuses on more general history regarding the need to communicate technical information to various audiences. Readers will want to read the rich histories of technical communication scholars in the field have written (Connors 1982; Longo 2000; Tebeaux 1996) for a fuller understanding of the discipline's roots. This chapter also identifies and challenges the assumption that technical communication is inherently objective and devoid of rhetoric.

Furthermore, because technology and science are interrelated, I borrow throughout this chapter (and book) from examples used by rhetoric and history of science scholars, but I avoid conflating the two areas and ask readers to consult additional Science and Technology Studies (STS) scholarship for a richer understanding of science as an ideologically mediated endeavor. Even though Johnson (1998) warned of "conflate[ing] science and technology as though they are of the same cloth" over a decade ago (p. 80), lay audiences still receive scientific and technical information in practically identical ways. We should avoid conflating science and technology, but we must also not ignore their similarities and obvious overlap. Because science also needs actors working together or, at least, interacting with experiments and research, both technology and science should be seen as comparable. After all, technology is often an application of science: In the case of the wireless, actors such as Marconi and others worked to harness the power of hertzian waves for commercial enterprise. Although expert-to-expert discourse might reveal major differences between science and technology based on a particular focus of applied vs. theoretical endeavors, communicators filter scientific and technical information to lay audiences in similar ways, making that delivery technical communication. This chapter focuses on that filtered technical communication.

Communicating scientific information is a type of technical communication, and science writing has rhetorical similarities to technical communication. Ultimately, if we could identify a muse for technical communication it would be technology itself. Therefore, there is a technology imperative inherent in technical communication, but technology as well as communication surrounding technology is mediated by the communities from which they come. The history of technical communication (or technical writing) demonstrates that it is both a rhetorical and socially constructed field.

1 Brief History of Technical Communication

Technical communication has an ancient tradition, and scholars have traced the field over the centuries. Arguably, technical communication can be considered the oldest form of communication. Although readers might not consider ancient writing, such as cave drawings, to be in the same category as computer manuals, such record keeping attempted to communicate information based on ancient humans' perceptions of their world(s). The drawings established knowledge and provided readers with guidance on hunting, gathering, and even astronomy. Oral cultures did not have the technology we have today, but they certainly carried out instrumentalist activities with primitive tools. Our contemporary bias is that technical communication is for highly technical endeavors, or, at least, technical communication deals with electronics. However, whether one needs to know how to find a sink hole in the desert or use the Internet to locate information on configuring new software, the information conveyed through discourse on those topics is technical communication. Ancient or contemporary technical communication, though, does not have to be didactic: Record keeping and advertising new technologies, for instance, are types of technical communication.

Although we cannot claim ancient to modern technical communication dealt with mutually exclusive activities, technical communication as we know it today appears to have been established during the Renaissance. Tebeaux (1996) * claims "Technical writing in the English Renaissance was, as it is today, writing for the world of work, except that daily life and work in the Renaissance were of one piece" (p. 3). I identify technical communication[1] as an activity beyond simply workplace communication, but Tebeaux's observation is typical of a general consensus that considers technical communication instruction as a practical job training endeavor. In addition to scientific communication, technical communication overlaps with business communication, and this marriage fails under a broad category of "Professional Communication." Technical communication is not universally defined that way, but much technical communication scholarship

---

* All quotations from Tebeaux's (1996) *The Emergence of a Tradition: Technical Writing in the English Renaissance,* 1475–1640 (from Baywood's Technical Communications Series, Series Editor: Charles H. Sides) are used with permission. Copyright © 1997, Baywood Publishing Company, Inc., Amityville, New York. All rights reserved.)

and pedagogy emphasizes its career-oriented disposition. In fact, Tebeaux broadens the scope of technical communication by claiming "technical writing is a basic form of humanistic expression" (p. 3). Machines are not made for the sake of making machines: Technologies are created to meet perceived human needs.

Miller's (1979) seminal text identified the field as germane to the goals of humanities education. In the article, Miller draws conclusions about both science and technology based on the similarity of their discourse. Technologists and scientists do not pursue their work as ends in and of themselves: Their work responds to social demands, but many go to great lengths to conceal the ideological frameworks surrounding science and technology. In fact, as Miller points out, the belief that technical communication has "little or no humanistic value is the result of a lingering but pervasive positivist view of science" (p. 610). The idea that technical communication transmits facts directly is a commitment to a positivistic paradigm, and one that falls apart when we become aware of the rhetorical nature of communication. However, much technical communication pedagogy today is positivistic, which "privileges expediency over critical technological awareness" and "promot[es] a skills-oriented, product-driven class" (Toscano 2011, p. 23). A commitment to positivism adumbrates the history of science and technology and causes one to assume that "[f]acts are self-evident entities existing out there in the real world—we have only to learn how to see them accurately or derive them logically" (Miller 1979, p. 612). History hides many scientific conflicts, causing the public to believe consensus is scientific fiat as opposed to rhetoric and consensus building.

The belief that discourse can directly communicate reality was held by Renaissance writers. Tebeaux (1996) traced English Renaissance technical communication and noticed its adherence to Ramist rhetoric, claiming "Ramist dialectic was likely the single greatest philosophical influence on changes in the visual presentation of information" (pp. 50–53). Although it is inaccurate to claim Ramist philosophy and positivism are identical because of positivism's strict adherence to observation and scientific methods, which would deny metaphysical explanations, "Ramists assumed that because One Reason ordered all things, Ramist method allowed the Truth of things to be opened and then compared" (Tebeaux 1996, p. 55). These English Renaissance writers assumed they could accurately communicate technical information because the world could be neatly ordered. Such an assumption ignores the observer's influence on ordering information. Under the Ramist paradigm, knowledge is independent and unchanging; it is universal. Even if this assumption has few disciples, it is a systemic attribute of technical communication. One of the main goals of technical communication is to limit reader (mis)interpretation. Unlike literature, which thrives on multiple interpretations and ambiguity, technical communication fails if it does not clearly articulate its goal. Renaissance authors were not privileging ambiguity:

> With the independence of the text, the role of person-to-person communication became unnecessary. The visual display of knowledge became monologic rather than dialogic, with the goal of text as single meaning emanating from the text to the mind of the reader. (p. 86)

These writers believed in a windowpane theory of communication.

Although limiting interpretation is a contemporary goal of technical communication, the field long ago recognized the fallacy of assuming technical communication follows a "windowpane" theory of transmission (Miller 1979, p. 613) or that technical communication is "an ethically and linguistically neutral activity" (Slack et al. 1993, p. 95). However, devotion to the assumption that "truth" is directly transferable from text to reader led to techniques advocated today. English Renaissance technical handbook authors rejected flowery presentations and "did not see the need for grandiloquence"; instead, they "emphasized directness in sentence structure, content, and presentation" (Tebeaux 1996, p. 134). Contemporarily, plain English advocates campaign for business, government, and technical texts to be accessible to average readers. The plain style is fundamental to technical communication instruction, and Tebeaux observes that "Renaissance writers" found it "appropriate for works written to be read and accessed rapidly for instructional purposes" (p. 134). Again, not all technical communication is didactic, but instruction is a common goal for technical communication. Tebeaux even traces technical writing in the English vernacular to Chaucer's Astrolab written in 1391, which was an instructional text he wrote for his son (p. 184). Also, Tebeaux credits Chaucer with "establish[ing] the description of describing a mechanism before presenting instructions for operating it, a method used in modern instruction and procedure manuals" (p. 185). Technical writers needed to communicate information about technologies to readers who had businesses to run, health care to practice, ships to sail, and even new technologies to create.

English Renaissance readers were hungry for information about new technologies and consumed handbooks on a variety of topics, which came about because of the social change from feudalism to a more market-oriented economy (Tebeaux 1996, p. 9). Tebeaux found "The answer to the demands of the middle-class readers for practical information appeared in the form of the handbook, the manual or guidebook, the Tudor and Stuart version of the self-help instruction book" (p. 10). Just as technology is a social development, Tebeaux argues "expanding literacy created a demand for books in the vernacular" (p. 4). The demand exists prior to the text's creation, and, as I show in the next chapter, demand comes before inventors try to create technologies. Tebeaux noted that a small market existed for books for experts, but a larger market existed "for books that provided more general instructions for less skilled readers" (p. 118). Most likely, the new middle class did not have the advantage of attending universities as aristocrats did. English Renaissance readers, as Tebeaux observes, could acquire an "[e]ducation, no longer limited to formal institutions such as grammar schools and universities," from "a wide variety of practical short courses and how-to manuals" (p. 177). These books were "for use as reference manuals rather than for thorough, sustained reading" (p. 36).

The reference quality of these English Renaissance technical works underscores a contemporary bias in defining technical communication: There is an idea the "true" technical communication relates to concrete subjects, which can be directly communicated through language. Such a bias ignores rhetoric and the ideology mediating technical communication and technology production. English Renaissance technical writing was a genre or, more accurately, multiple genres doing cultural work. In this case, these genres reflected the needs of a diverse, diffused society dealing with common technologies brought about by changing social (and economic) conditions. English society (as well as the rest of Europe) was increasing its "global" reach and would dominate the seas by the end of the Renaissance (circa 1640) for the next three centuries. It would have been impractical to implement a national compulsory education program to promote learning in the practical arts—the subjects of the English Renaissance handbooks (Tebeaux 1996, p. 11)—but technological advancement and, concurrently, knowledge of these advancements flourished. Tebeaux noted "Writing and printing helped disseminate knowledge to an ever widening circle of newly literate readers who came to depend on the text rather than orally dominated instruction for usable information" (p. 177).

The exchange of ideas is a surface reason that technology flourished. The commitment to exchanging ideas, improving upon extant materials, systems, and procedures fueled technological advancement. I will show in Chap. 3 that Marconi's wireless developed, in part, because of exchanging information on the science of electromagnetic waves and ways to harness their power, but I want to point out that Tebeaux (1996) noticed a similar relationship between technical communication and technology advancement mediated by English Renaissance culture: "[T]echnology developed in response to need, and technology and literacy seemed to nurture one another" (p. 133). Also, "[t]he emergence of a sophisticated style indicates the need for such a written discourse to textualize advancing information to communicate improvements in technology" (Tebeaux 1996, p. 133). And readers needed to believe in the validity of "such a written discourse" in order to accept the handbook's authority and perception that it communicated truth.

Therefore, both readers and writers of English Renaissance technical books had to assume language could transmit accurate perceptions of the natural (as in the case of science) and mechanical (as in the case of technology) worlds. Tebeaux argued "Renaissance technical writers, without question, had a universalist view of language" (pp. 244–245), but contemporary technical communication textbooks have such a view of language. Some technical communication scholars question the ability of language to communicate *objective* truth (Miller 1979; Dobrin 1985; Rutter 1997; Samuels 1985), but the field has a desire to be objective. No where in the scholarship is this view supported with greater conviction than in the work of Patrick Moore. Moore's (1996) discussion about why the field of technical communication should embrace instrumentalist discourse counters subjective or rhetorical perspectives, which can be summarized by his following observation:

> In many kinds of technical communication, language is denotative because it is standardized; that is, a group of people agree—sometimes in writing—to call a certain thing by one name, execute a procedure in one way, give a certain object specific measurements, or give a system certain specific functions. (p. 108)

Readers may sympathize with Moore's view when considering, for instance, the problems that could arise if a nuclear power plant operated on ambiguous, unclear standards. Moore argues that standards enforce continuity of community-defined commitments that "many people must agree to change," thus, "unit[ing] many people into one team that works in a coordinated way to solve the large, complex problems of organizations and nations" (p. 109). Moore does acknowledge that objectivity is "impossible" (p. 108) and that adhering to a "windowpane" view of language is invalid, but he also cautions against the belief that reality cannot be transmitted because "[i]n many technological situations, a rigid, one-to-one correspondence is required between the signifier and the signified or else someone could die" (p. 110). He uses a medical example to prove his point, but a mechanical one could be that readers would most likely not feel safe and, therefore, refuse to fly in airplanes that were maintained by mechanics who felt tightening all the bolts was optional.

However, technical communication is still a rhetorical endeavor even with recognizing instrumentalist needs for communicating specific, unambiguous information. Moore (1996) assumes rhetoric is simply persuasion and warns that "[f]ew people have the time, the need, or the inclination to be persuaded about many activities in their lives. In many situations, people need closure—not more interpretations, more analysis, or more discussion" (p. 115). He is absolutely correct in observing "that people must use language to get things done—to execute physical tasks within narrow financial, temporal, and other constraints" (p. 115). What complicates the positivistic paradigm, under which Moore argues, though, is that through standardization, professionalization, and other types of socialization, members of a culture (or subculture) have already been *persuaded*. Scientists are committed to the scientific method and would not consider unrepeatable tests valid knowledge construction. Engineers are also committed to testing and would not consider allowing the public to occupy a building without performing various, standardized integrity tests. In fact, the above commitments are held deeply enough that legal codes, established by consensus, supplement scientific and technical authorities. As later chapters show, however, technical communication occurs even when there is little or no consensus on the validity of particular scientific and technical apparatus. Not only do scientific and technical authorities develop and adhere to standards, but the cultures in which these authorities exist are committed to believing technical and scientific solutions should be pursued. Again, the overall culture demands its needs be met and accepts technological advances (regardless of whether or not the acceptance is universal). Without a commitment to recognizing the validity of scientific and technical solutions, English Renaissance technical communication (or any group interested in practical knowledge) would not have developed. In other words, technical writers would not have a reason to fulfill the demands of audiences devoted to their genres.

At a basic level, technical communication cannot come to be without a social commitment to technology production. After all, "technical writing adapts technology to the user" (Dobrin 1985, p. 247). I wish to extend Dobrin's definition by revising it to be *technical communication, a practice mediated by culture(s), involves discourse that informs audiences about technology.* The above definition does not limit technical communication to user-specific genres (e.g., manuals) but attempts to capture all the discourses surrounding technology. Before discussing how past technical communication relates to modern technical communication discourse communities, I will turn to a brief discussion on how a pervasive cultural bias regarding the pursuit of perfection through technology influenced attributes about technology and served as a justification for Western technological advancement.

1.1 Pursuit of the Industrial Arts

The goal of this section is to argue that the belief in objectivity, truth, or a monolithic organizing principle mediates both technological advancement and technical communication. Regardless of any instrumentalist need to communicate information unambiguously, rhetoric is a priori to discourse. That does not mean technical communication is reduced to the common definition of rhetoric, which assumes "rhetoric" is simply empty communication. On the contrary, it means rhetoric is a substantial governing principle of a community's discourse practices. In other words, the ideas inherent in discourse surrounding technology are mediated by social commitment to the efficacy of technology. Demand for technology exists before its creation, and belief that language can transmit this knowledge exists before discourse. Furthermore, members of a culture hold predisposed notions regarding technology that influence attitudes about technical communication. The ideology that mediates technological development is entwined with the ideology mediating technical communication. I will now turn to an example from David F. Noble that further develops this thesis by highlighting how cultural beliefs influenced both a commitment to technological development and discourse communicating the value of particular technological pursuits.

Noble (1999) identifies the religion of technology as a commitment to pursue technology in order to regain "mankind's prelapsarian powers," which were lost after the fall of man (p. 17). Noble traces this idea over the last millennium throughout his book and observes that scientists and engineers appeared to pursue "the mechanical arts" in order to get closer to their idea of god's perfection (15). Such an ideology would be impossible without cultural support, but the entire culture does not have to engage in what Noble demonstrates as an elite activity. The early Middle Ages saw a change in how society, specifically the elites, viewed technology:

> Technology came to be identified more closely with both lost perfection and the possibility of renewed perfection, and the advance of the arts took on new significance, not only as evidence of grace, but as a means of preparation for, and a sure sign of, imminent salvation. (Noble 1999, p. 12)

Just as Tebeaux notes that a change from feudalism to a market oriented economic system—a social change—led to demand for new handbooks, Noble identifies the ideological change as a "new millenarian mentality" that "situated the process of recovery [from the Fall according to Christian mythology] in the context of human history," meaning humans were able to engage in "the pursuit of renewed perfection" (p. 22). Salvation was no longer limited to living a pious life; instead, cultural elites pursued new advancements to get back to godlike perfection, and this agenda "was to have enormous and enduring influence upon the European psyche, and it encouraged as never before the ideological wedding of technology and transcendence" (Noble 1999, p. 22). Noble traces the belief in transcendence through technology to contemporary technology, such as Artificial Intelligence (AI), which "AI practitioner Daniel Crevier" claims "is consistent with Christian belief in resurrection and immortality" (p. 160).

The next Chapter goes into more detail about social constructions of technology, including the religion of technology, but Noble (1999) offers examples of technical communication, although he does not refer to discourse surrounding technology as technical communication, being mediated by this elite millenarian ideology throughout his book. Therefore, the religion of technology is also a rhetoric of technology because the technologies engineers pursued fit with the goals the elite societies (e.g., The Royal Society) had in creating a more perfect world. Noble defined "the religion of technology" as humanity's pursuit to discover "prelapsarian powers" (pp. 16–17). Although it is impossible to claim all elite members of the Renaissance and later time periods believed they could return to the days of perfection, the assumed state in the Garden of Eden, Noble identified major figures in science and technology who pursued their work in order to restore humanity's assumed perfection. Of course, most of the figures Noble discusses are not average citizens: They are the elite, major figures of Western civilization. However, their pursuit would be impossible without cultural support. These elite figures communicated their scientific and technological discourse, which was laden with concerns of prelapsarian power. The dominant ideology mediates their technical communication.

Noble (1999) argues that "[p]erhaps more than anyone else before or since, "[Francis] Bacon defined the Western project of modern technology" (pp. 48–49). Additionally, "Bacon," who was also a product of his time period, "viewed practical knowledge of the arts as the key to the advancement of knowledge in general, and used the mechanical arts as the model for the reform of natural philosophy" (p. 49). Bacon and other Restoration scientists/philosophers, as Charles Webster observed, "were concerned to give a vivid impression of the great power sacrificed at the Fall, in order to galvanize their contemporaries into an effort to restore the primitive condition" (as cited in Noble 1999, p. 46). The pursuit of regaining prelapsarian *perfection* was tied to the supposed 1,000 year reign of Christ, and "[t]his unprecedented millenarian milieu decisively and indelibly shaped the dynamic Western conception of technology. It encouraged a new lordly attitude toward nature, reflecting the anticipated restoration of Edenic dominion" (p. 49).

Although Noble's (1999) purpose is to explain how technological creation was mediated by the time period's millenarian ideology, he notes technical communication situations in which Bacon and others engaged. Bacon had to convince contemporaries that "their elite disdain for the useful arts" was misguided, and he continued the tradition of uniting "the mechanical and liberal arts" as equally valid pursuits (Noble 1999, pp. 49–50). Bacon certainly had other aims for pursuing scientific knowledge, but, as "Paolo Rossi described it, Bacon's overriding aim 'was to redeem man from original sin and reinstate him in his prelapsarian power over all created things'"(as cited in Noble 1999, p. 50). Bacon communicated his goals through his works, which I am arguing is a form of rhetorically laden technical communication, and "[he] was explicit and insistent about the perfectionist purpose behind his advocacy of the useful arts" in his famous work *The Great Instauration* (p. 50). He was not writing to the masses or the dilettante merchants, as were the writers of the English Renaissance handbooks Tebeaux identifies; instead, he wrote to convince the establishment—philosophers and scientists—that pursuing the useful arts fit the "divine plan" of restoring "mankind's original powers" (Noble 1999, p. 51). Bacon's technical communication shows it is mediated by dominant ideology and aimed at the elite who controlled scientific and technical apparatus.

Noble (1999) traces Bacon's influence in his technical philosophical works, noting "[l]argely through the enormous and enduring influence of Francis Bacon, the medieval identification of technology with transcendence now informed the emergent mentality of modernity" (p. 53). Another connection Bacon has to modern technical communication relates to our contemporary bias regarding the proper role of technical communication: It has a didactic quality. Rossi (1968), an important biographer of Bacon, noted that Bacon's works promoted reforming science education:

> traditional learning must be replaced by the cult of nature so as to re-establish the contact between man and reality; collections of facts are a means of study, an instrument for scientific research and not objects of pleasure and curiosity. (p. 9)

Bacon was not the only one advocating this change, and Rossi argued the point that the view was a cultural construction, for

> Indeed Bacon was voicing the general opinion of his age, defining some of its essential demands, when he strove to rehabilitate the mechanical arts, denounced the sterility of Scholastic logic, and planned a history of arts and sciences to serve as foundation for the reform of knowledge and of the very existence of mankind. (p. 9)

Rossi also argued that "[t]he basic themes of Bacon's philosophy…were directed at specific objectives and may be ascribed to a definite phase of culture" (p. 11). Additionally, Bacon established rules for communicating scientific knowledge consistently by building on the philosophers of the past.

Of course, drawing on such philosophers as Aristotle, Plato, and Socrates necessarily means any scholar would have assumptions regarding rhetoric. Rossi (1968) noted Bacon believed that rhetoric's "function is to extend the empire of reason and defend it against every onslaught, even if in a different field and with different tools" (p. 178). Bacon's own definition clearly applies to technical communication because he did not consider rhetoric as simply flowery language;

instead, "the duty and office of rhetoric is to apply reason to imagination for the better moving of the will" (as cited in Rossi 1968, p. 180). Bacon's plan, therefore, is to convince audiences that nature can be directly conceived from an observer's prose. According to Rossi, Bacon assumed "[m]ost mechanical arts originate from observations of nature or natural phenomena—for indeed nature is 'art's mirror'" and include "reproductions of the spectrum, distillation, artificial thunder and lightning, etc." (p. 155). Although some contemporary technical communicators might assume a windowpane or direct reflection of nature is non-rhetorical, Bacon saw proper rhetoric as "stable and unimpassioned....hav[ing] a specific function which is not to 'delight' but to attain certain predetermined ends" (Rossi 1968, p. 185). Bacon's legacy pervades the Royal Society, which had the purpose to "free the present age [c. 1667] from errors of the past," (Rossi 1968, p. 184). Bacon wanted scientific discourse to be an agent of "purifying the intellect," and he believed "the path of righteousness...require[d] the assistance of rhetoric" in order not to excite passions. (Rossi 1968, p. 184). The rhetorical tradition in technical communication has a long history and, as the above discussion of Bacon makes clear, is directly concerned with audience.

Although Bacon wrote for the elite, Tebeaux (1996) observed that credit attributed to him with regard to establishing "plain style" as the preferred discourse of science and technology was inaccurate because "plain style" was a cultural product. Tebeaux claimed "Bacon's principle contribution was a theory of applied discourse—given credibility by his political and intellectual stature that provided a rationale for a style already used extensively in various types of functional discourse" (p. 164). Furthermore, continuing the connection to contemporary technical communication pedagogy "Bacon believed that selection of ideas requires that writers decide what to include and what to reject" (Tebeaux 1996, p. 164). Bacon's legacy goes beyond technical communication and can be seen in a familiar strategy advocated by Composition scholars stressing audience, purpose, and context with their students: "Bacon believed that the order and arrangement of all discourse, both scientific (instructional) and philosophical, should be determined by audience, purpose, and occasion" (Tebeaux 1996, p. 165).

All communication—not just technical communication—must adhere to audience expectations to be effective. Bacon understood that on a surface level, but he also fit audience expectations on a deeper level because his content or project matched millenarian ideology. Noble (1999) claimed the "millenarian mentality [of the seventeenth century] gave formative shape to the milieu of modern science" (p. 57). Noble identified more consequences of this ideology: "Like Bacon, the founders of the new scientific academies also tended to view science as technology, as a philosophical enterprise inextricably bound up in both method and purpose with the useful arts" (p. 57). Of course, these organizations, such as the Royal Society of London and various continental ones, "were utterly imbued with the Baconian spirit of usefulness" (Noble 1996, p. 59). Additionally, Margaret Jacob links these elite scientists together and noted "[a]lmost every important seventeenth century English scientist or promoter of science from Robert Boyle to Isaac Newton believed in the approaching millennium" (as cited in Noble 1996, p. 59). These scientists, engineers, and businessmen pursued their work and, in turn, communicated to each other. This communication, mediated by millenarian ideology, was technical communication laden with millenarian references and assumed to be guided by divine intervention (Noble 1996, p. 67).

The next section works from Bacon and English Renaissance plain style to demonstrate how that time period's technical communication contains a didactic bias that continues to mediate modern technical communication.

1.2 Didacticism in Modern Technical Communication

Although modern technical communication shares attributes similar to Renaissance and older forms of technical communication, the field has been defined in narrower academic terms over the last century. Tebeaux (1996) uses Dobrin's definition of technical writing to emphasize the practice's commitment to *technology* and its goal of efficiency: "If we define technical writing as that kind of writing which adapts technology to the user, then we can see that process description and instructions were naturally inclined to unadorned statement and direct presentation" (pp. 168–169). Plain language, whether Renaissance or contemporary, is a quintessential goal for technical communicators. Of course, plain language is not just a goal of technical communication, but it is a necessary feature of effective technical communication carried out in procedures, manuals, reports, and the like. Therefore, we can define technical communication broadly in a historical sense as plain writing (or communication) on technical subjects. Tebeaux's analysis of how-to handbooks easily fits the above definition, but many today would not immediately consider how-to guides (e.g., *The Dummies* guides series) or general technology reference as "technical communication." The reason is a century old academic bias in viewing technical communication as professionalized discourse learned in the academy and on the job.

Modern technical communication, which I am bracketing as stemming from disciplines since the Scientific Revolution, relates to expert-to-expert discourse in a "pure" form. In other words, highly technical information dissemination, in theory, is supposed to be a direct communication of technical information among members of a discourse community. Scholars have pointed out that such an assumption of direct communication among experts is inaccurate (Latour 1987; Latour and Woolgar 1979; Winsor 2002), but many engineers, IT professionals, and scientists claim that they communicate directly to experts who understand what is being discussed. Of course, this is not a universal outlook, but it is a commonly enough held assumption that it becomes a reality regarding technical communication, thus, affecting how we define the subject across disciplines. Also, consumer-related documentation has been a major form of technical communication since the end of World War II (Connors 1982, p. 342).

Contemporarily, many view consumer-related discourse (e.g., manuals) as technical communication, but other lay-audience-specific texts (e.g., journalistic reviews), which, essentially, adapt technology to the user, are not included in the academic definition. A major reason for not including general information on technology as "technical communication" has to do with historical discourse surrounding science and technology. As Tebeaux (1996) observes, "the 'plain style' advocated by the scientists of the Post-Restoration Royal Society"—the elite involved in expert-to-expert discourse—was not the first time technical writing exhibited an efficient style (p. 138). Certainly, though, organizations like the Royal Society and Royal Institute sustained technical communication because their goals were to advance scientific and technical knowledge. New ideas had to be communicated.

Unlike the handbooks Tebeaux analyzed, which appealed to non-scholastic learners, textbooks fulfilled post-Renaissance demands of universities. Longo (2000) observed that, as more technical-oriented universities were established and science and technological knowledge spread, textbooks began to appear in the seventeenth century (p. 24). These textbooks were distinct from previous "books of secrets" that advocated what we would today call pseudoscience (Longo 2000, p. 24). Additionally, Longo notes the "tension between private and public science" threatened religious dominance (p. 29). It is important to the definition of technical communication to discuss the problems the religious community had with science and, as Noble (1999) pointed out, the industrial arts. The shift from a theocratic-central view of *truth* to a material view is a cultural phenomenon. Basically, such a shift is a revolution that replaces the time period's dominant ideology. Without this shift, science would not have existed as an objective discipline; instead, observations would have had to have been reconciled with religious doctrine, a perhaps procrustean endeavor. As in the case of Galileo, seventeenth-century papal authority condemned his pro-Copernican view of heliocentrism, which violated established dogma of geocentrism. It took the Vatican nearly 400 years to exonerate Galileo in 1992. Four hundred years is an eternity in terms of scientific and technological advancement, but the Vatican (as well as other religious institutions) still adapted to later scientific revolutions. As Noble pointed out, without the church sanctioning "[t]he dynamic project of Western technology," inventors could not have pursued advancements openly (p. 9). Therefore, without social support (both positive as in the case of patrons and negative as in not condemning scientists and technologists) exchanging ideas via any form of technical communication would have been difficult if not impossible.

And how did scientists and religious authorities compromise? Instead of openly challenging religious authority, which would risk severe punishment, science incorporated itself into Christian mythology. Bacon, through discourse, "set the groundwork for persuading other intellectuals" to pursue "scientific observation" (Longo 2000, p. 41). Additionally, "[b]ecause Bacon's scientific method was grounded in biblical teachings, experimenters could explore natural phenomena without fearing that they blasphemed by seeking to upset God's plan on earth"

(Longo 2000, p. 41). Bacon and others advanced their cause by aligning them-selves with established religious doctrine and made the work seem as if they were doing the work of the divine. Bacon's practice, which led to the formation of the Royal Society, "was a method for achieving social dominance for science through the promise of increased security and an improved human condition"; furthermore, "[t]echnical language would become the lingua franca of this scientific society and its institutions" (Longo 2000, p. 43). As I show in Chap. 3 how Marconi's tech-nical discourse adhered to his time period's ideology, the technical discourse of Bacon and his contemporaries was mediated by the socially constructed imperative to have science fit narratives concerning god's role in the endeavor. Technical communication is always from a priori rhetoric, including assumed or unconscious social agendas.

Even though technical communication discourse was mediated by religious doctrine, its style began to show its distinction from religious and non-technical texts prior to Bacon's influence. As information exploded during the English Renaissance, the printing press fulfilled the demand of producing texts for various audiences. No longer could oral methods of knowledge dissemination satisfy the society's technical communication needs (Tebeaux 1996, pp. 176–177). The *function* of technical texts influenced the *form* and, of course, style of delivery: "Unlike ornamentation and illumination techniques used in books of prayers, devotions, and poetry, visual design in technical books was functional rather than ornamental, communicative rather than impressionistic" (Tebeaux 1996, p. 175). Not only do we see form following function, but we see culture mediating form. Tebeaux argued "it was not science with its attempt to capture truth precisely in words which led to the rise of plain style but an increasingly literate public that needed books written in spoken English for self-enhancement" (p. 169). Just as English Renaissance audiences demanded texts in the English vernacular (Tebeaux 1996, p. 14), any culture's demands mediate knowledge production and commu-nication of that knowledge. Tebeaux noted that "[a]llusions to God declined" as the Renaissance progressed, but "early Renaissance technical writing recognized the role of God's grace in all work and asked God's blessing on all forms of human endeavor" (p. 13). Such a commitment reflects hegemonic forces in the culture. For the Renaissance and even the Scientific Revolution, the church was an influential hegemon affecting both scientific and technical pursuits as well as the discourse surrounding these pursuits.

Although today overt religious invocations are absent from technical texts, Noble argued a specific religious affinity, millenarianism, guided technological production in Western society. Noble's position is debatable, but such a debate/ refutation is beyond the scope of this book. However, the next section advocates that another hegemonic force, the academy, which is not mutually exclusive from millenarianism, influenced technical communication and, in turn, influenced how we define technical communication.

1.3 Technical Communication in the Twentieth Century

Technical communication education and practice as known contemporarily emerges from an academic need, which, ultimately, supports the culture's hegemonic structure. As I demonstrate in later chapters, Western technology in the twentieth century adhered to militarization goals. Even though the military or, more accurately, the hegemonic view that a nation ought to pursue technologies to improve military dominance influenced the ways inventors like Marconi discussed the benefits of new advancements. Additionally, an emerging academic discipline in the early twentieth century influenced perspectives about communicating technical information. Longo (2000) traced the development of technical writing over centuries and identifies many influences, but her thesis and, therefore, project is guided by cultural study (pp. 19–20). Although I make no attempt to recreate her rich exploration regarding histories of science, business, and technical communication (she uses the term "writing"), I do rely on important observations she makes about technical communication in the twentieth century for my definition of technical communication—all discourse surrounding technology. Longo recognized technical writing's impact and purpose in a technocratic mass society: "Technical writing has been used to track the activities of people and machines, with the goal of assigning value to those activities. Technical writing is the control mechanism of scientific and technical knowledge production" (p. 2). What I want to make clear is that technical communication *assigns* value both in professional and popular communication situations. I will now explain how technical communication's academic location in the early twentieth century fits my more expansive definition.

The twentieth century provides evidence of changing the legitimate mouth pieces of technical communication. Longo (2000) identified T. A. Rickard as being the first to articulate technical writing's practice "for assigning value to knowledge" (p. xiii). Also, in Robert Connors seminal work on the history of technical writing, he claimed "[t]he first notable textbook devoted to technical writing was…Rickard's *A Guide to Technical Writing*" (p. 332). Rickard, a former mining engineer, wrote the textbook in 1908 and argued "technical language is a coinage that engineers should contribute cheerfully to the general fund of scientific knowledge for the betterment of living conditions" (Longo 2000, p. 22). Because "textbooks are cultural artifacts participating in knowledge/power systems" (Longo 2000, p. 24), we can observe how culture mediates these (and other) texts. Rickard wrote his textbook for mining and metallurgy students, and, along with advocating that "scientific and technical language should be purified" (Longo 2000, p. 65), he described proper engineers and technical writers as men adhering to positivistic, objective goals (Longo 2000, pp. 69–70). Equating science and technology as reflecting truth is a modernist commitment explained further in Chap. 5 where I discuss F. T. Marinetti's technophilia. Although one cannot claim all universities used Rickard's book, Robert Connors argued it was a popular textbook that "sold well and was adopted at a number of schools" (p. 332), which suggests it fit in with academic culture. The book, at least, reflected a major belief about discourse in engineering and science fields.

What is important to note about the academy is that its mission, in part, was to develop technological solutions to meet social demands. Rickard warned that if engineers continued to discuss scientific and technical knowledge without adhering to acceptable standards of "technical language, engineers threatened science and the survival of the human species" (Longo 2000, p. 65). Not only does such a view conform to the time period's pro-technology-solution ideology, but Rickard's own words regarding technical communication are infused with tropes of progress:

> language is a factor in the evolution of the race and an instrument that works for ethical progress. It is a gift most truly divine, which should be cherished as the ladder that has permitted of an ascent from the most humble beginnings and leads to the heights of a loftier destiny. (as cited in Longo 2000, p. 64)

Rickard definitely follows the rhetorical tradition of Bacon and Noble's argument about the Royal Society's members attempting to recover the prelapsarian Eden. What makes Rickard's technical communication rhetorical is not just his metaphors and allusions to technological advancement leading to greatness but his audience's predisposed belief, mediated by modernist ideology, that science and technology *offer* solutions. Without a commitment to the socially constructed perspective that technological advancement is desirable, Rickard's words would never receive any attention. Technical communication, as with all discourse, does cultural work by reproducing cultural values and attitudes that mediate the readers' or listeners' practices. In the specific case of academic technical communication, readers' practices involved pursuing technological advancement.

Of course, Rickard and others were communicating to advanced audiences. The "average" person in an industrialized culture is immersed in technocratic politics regardless of whether he or she knows it, but that immersion does not mean this citizen understands expert-level discourse. As I show in Chap. 3, technical communication, even in technical forums, is audience specific. The communicator's choice of language (e.g., highly technical vs. semi-technical) requires employing rhetorical strategies to be effective. However, not all technical information aimed at lay audiences is considered technical communication, which this chapter refutes. Before explaining how different rhetorical strategies affect the perception of technical communication, I now turn back to technical communication professionalization during the twentieth century. Technical communication's academic "regulation" continued, but the individuals allowed to reproduce technical knowledge changed because of social needs.

Longo (2000) observed that textbooks, which communicate the goals of technical communication in the twentieth century traditionally serve two functions: (1) they "compile important knowledge about a specific topic in the encyclopedic tradition," and (2) "they also attempt to include recipe-like accounts of how to make the textbook information useful in some practical way in the tradition of the Hermetic books of secrets" (p. 70). Although lay audiences and even enthusiasts might enjoy encyclopedic information about science and technology, many early twentieth-century technologies would be difficult for "backyard inventors" to replicate. After all, many advancements were military applications. The market for information on such technology as wireless experiments existed, and, especially for children, popular fictional figures had adventures with new technology. One popular children's series about a boy inventor, Tom Swift, has the young Swift building a wireless radio out of parts of a crashed airplane (Appleton 1911). But technical communication was not just for didactic reasons. Information in various forms—newspapers, public demonstrations, and, eventually, radio—met the demands of a technocratic society, one that expected and wanted to absorb information about new technologies. As Chap. 4 demonstrates, articles about new technology filled popular newspapers and journals. Still, though, another forum reproduced technical information and was mediated by the academy: the work of the professional technical writer.

1.4 Post-WWII Rise of Technical Communication

Longo (2000) found that "Specialist technical writers did not appear until after World War II, when some organizations split some communication functions from research and development functions in order to make technology development more efficient" (p. 2). Although one can read the first half of the twentieth century as having an ideology supporting ever increasing efficiency standards, there are other perspectives that do not contradict that view. Longo argued that technical communication's status was mediated by industrial society's commitment to management science, specifically the theories of "Frederick Taylor's system of scientific management," which "[were] extended to the control of production and management itself" (p. 99). In Longo's own observation, though, we can see the root of industrial nations' commitment to efficiency: militarization, which I will show to be the overarching principle of the first half of the twentieth century. Longo identified one of the key informers in American business, Edward D. Jones, as articulating the apparent successful link between German military culture and business success (p. 108). In order to satisfy industrial nations' goals of advancement, nations had to pursue (or, in the case of Britain, continue) colonization, which required militaries.

World Wars I and II were byproducts of industrial nations' attempts to gain and maintain dominance. The end result of these wars was the development of weapons of mass destruction. Although the atomic bomb had the immediate effect of convincing the Japanese to surrender, its development followed the hegemonic desire (of powerful colonial empires and those aspiring to empire) to steadily increase a nation's ability to destroy, an effect of militarization. Even today we can see that militarization spawned everyday technologies from satellite television to the Internet. The twentieth century was committed to war and preparation for war. And American institutions, including schools and businesses, followed the overarching ideology of militarization, thus, developing technologies to satisfy hegemonic demands. To turn out the weapons, vehicles, and ideas (science) that followed such an ideology, businesses had to be efficient and centrally controlled like the military. Longo identified this relationship by claiming, "[m]anagement engineering...looked to military history for models of efficient large-scale organization and technical writing provided the control mechanism for these organizations" (p. 109). What happens, eventually, is the communication demands will be too much for scientists and engineers, requiring them to have support.

This support came in the form of professionals devoted to technical communication. Again, following Fordist/Taylorist practices of increasing efficiency by reducing the number of tasks a worker has, technical writers take on the "burden" of the communication of engineers and scientists. Longo (2000) claimed technical communication moved out of the hands of scientists and engineers, and "the role of technical writer was transformed into lower-paid help to relieve higher-paid engineers and scientists of the burden for generating communications that stabilized the dominance of scientific knowledge in our industrialized society" (p. 149). After World War II, technical communication increasingly begins to be taught in English departments, and organizations supporting professional technical writers begin to emerge (Longo 2000, pp. 150–151). Therefore, English departments "become productive members of society" by assisting knowledge transmission of "the science and engineering departments," so they may "better use their time developing scientific knowledge and technological applications than teaching engineering support staff to write" (Longo 2000, p. 150). Even today technical communication is largely housed in English Departments (Yeats and Thompson 2010), and, as I discuss later, the academic focus of technical communication privileges consumer-oriented, didactic forms of technical documents as opposed to popular transmissions.

Furthermore, any cursory look at technical communication textbooks reveals technical communication is predominantly a workplace endeavor—professionals communicate technical information while "on the job." Certainly, technical communication is a workplace practice: Even if an employee in a professional setting does not have the job title "technical writer/editor," he or she will most likely communicate technical information. But technical communication, which acclimates audiences to information about technology, happens in a variety of situations through different media. Because technology and, therefore, communication about technology is mediated by culture, audiences *learn* about technology in ways broader than assumed, dominant discourse advocated by academic practices. Besides the traditional reports and consumer materials associated with the definition of technical communication, technical knowledge is disseminated in a variety of ways. The next section discusses how technical information reaches audiences through expert-oriented discourse, journalism, science fiction, and culture itself. After all, as McLuhan (1964) observed, technologies communicate implied messages of a culture (p. 8).

2   Knowledge Dissemination Through Various Technical Communication Strategies

Up to this point, I have focused on how technical communication advanced through practices related specifically to science and technological advancement/development. Even though we can trace the cultural work of technical communication, its adherence to cultural ideology, through the works of Bacon, Rickard, and others, technical communication happens outside of expert-to-expert discourse. Although a professional "technical writer" is often a non-expert, with the exception of freelancers, perhaps, technical writers are industry specific. For instance, an engineering firm may have a technical writer to respond to requests for proposals, a software company may hire a technical writer to document the latest release of a program, and a consulting firm may hire a technical writer to prepare compliance documents for lawyers (often known as professional writing). The above technical writers may communicate to lay audiences, the most likely case for software documentation, but they produce more technical communication for specific semi- and highly technical audiences. However, lay audiences receive technical communication from outside the marketplace, and the communication is mediated by socially constructed values and a priori rhetoric—meaning is already constructed for members of a culture.

Knowledge advances because discourse communities communicate with their members, but, if experts only communicated with other experts, knowledge would not diffuse to other audiences. Technical and scientific information pass to general audiences through a variety of ways: popular press articles, textbooks, media, and even science fiction. I will address science fiction later, but in this section I trace knowledge dissemination from expert to lay audiences. Although it is not always recognized as such, non-instrumentalist discourse about science and technology is technical communication. If technical communication acclimates users to technology, defined above by Dobrin (1985, p. 247), we cannot ignore discourse that introduces users (and non-users) to technology. For instance, before one begins to program a mobile communication device, some type of communication made audiences aware that the possibility for instant communication existed. Likewise, before a team of engineers produce a more efficient internal combustion engine, some type of communication made them aware of techniques to employ to get the engine to run longer on a gallon of gasoline and, more importantly, that a market existed for a more fuel efficient engine.

Part of the process of constructing knowledge revolves around naming. Naming a new concept gives power to the knowledge by capturing the essence of the meaning efficiently in a word or phrase. Naming is a way of black boxing ideas so knowledge producers may continue to build upon past ideas. As Tebeaux (1996) observed in her analysis of English Renaissance technical writing, "[k]nowledge, in order to grow, had to be able to have meaning in non-oral textual forms that could be preserved and reproduced graphically for rapid access" (p. 240). For the expert, having a name for an idea captures not only the essence of the knowledge but the history behind the development of that knowledge. Experts do not have to keep (re)describing the development of a particular knowledge concept in their discourse. Advancement, technical and scientific, is always based on the ideas of past knowledge. In the next chapter I critique the narrative claiming technological change is inherently progress(ive), but I want to mention here that the narrative of technical progress is, itself, technical communication. Progress narratives are mediated by the dominant cultural belief that technology always provides better solutions. Such a value is more accurate for the contemporary consumerist culture of America, and it shows that consumers are buying products—lots of products—to fulfill their desires. Additionally, the work of Kuhn (1962) is important to mention because he shows that scientists communicate their findings, which is technical discourse, and, eventually, scientific revolutions take place when major changes to scientific thought occur. These changes only come about when a majority or plurality accepts new understandings of science. Technical communication, in a variety of forms, disseminates these new understandings.

Because knowledge is community based, we can claim knowledge "speaks" to intersubjective truth. Of course, there are *objective* ways of codifying knowledge, and engineers and scientists do create concrete material products, but the standards by which they create materials are community based. Additionally, these communities have authority for establishing knowledge that is accepted by audiences outside the community. The general public may take the word of an established expert on a technical subject even if the expert is not communicating to the public using highly technical descriptions and jargon. For instance, although this subject is not without controversy, the general public may believe scientists who warn about the effects of greenhouse gases on the atmosphere even if the public does not fully understand how these gases (or which specific gases) lead to global warming. The general public may know little or even nothing about the ways scientists discovered the effects of greenhouse gas-emissions on the climate, and they might not understand the peer-review process scientists go through to have their communities legitimize the knowledge produced, but scientists and technologists are considered knowledge authorities.

The discourse from a knowledge authority or the invocation of a knowledge authority is a rhetorical strategy, an appeal of ethos. The simple phrases "experts agree" or "Dr. Winslow urges" affect an audience's reception of the message. Journalists consult experts of various fields frequently when addressing technical topics, but their discourse does not mimic expert-to-expert discourse like that of an academic journal. Similarly, introductory textbooks communicate technical information to general or, in some cases, semi-technical audiences (i.e., junior- and senior-level college textbooks). Textbooks and mass media (journalistic) discourse do not necessarily communicate the process of knowledge construction, but audiences may absorb the information authors present without having an advanced degree in the subject. Just as a journalist may cite an expert in the field of climatology to discuss issues related to global warming, textbook authors, usually experts in a field, are understood as authorities, thus, legitimizing the knowledge presented.

Even when non-expert discourse inaccurately or incompletely represents technical information, the discourse is still technical communication. "Truth" or objectivity might be a goal for technical communication, but it is not necessarily an inherent feature. An interesting example of this is in the reporting of condom effectiveness based on a June 2001 National Institutes of Health (NIH) review of 138 peer-reviewed articles. The NIH panel reviewed a selection of articles published from 1969 to 2001 and concluded that they did not have enough information to verify or dispute condom effectiveness against STDs with the exception of heterosexual HIV and gonorrhea transmission. The 49-page NIH report was "summarized" into a

1-page (542-word) article on CNN.com with the title "Condom report questions STD protection" (Feig 2001). Such a title, however, is misleading because the NIH report concluded there was *inconclusive evidence to refute or justify condom effectiveness*. The distinction is subtle, but the title Feig uses assumes the government panel concludes condoms are ineffective. The title, which could have been chosen consciously or unconsciously, predisposes the reader to "question" condom effectiveness. Therefore, the title's rhetoric may cause readers to believe the NIH study, produced by knowledge authorities, claims condoms are *not* effective. Even inaccurate technical communication can be effective, influential, or just convincing. The CNN.com article is a type of technical communication because Feig reports, selectively, about scientific and technical methods discussed in the NIH report. The article, however, does not have the same peer-reviewed process academic journal articles have. One question to ask is do audiences receiving technical information in a general forum care about the process of knowledge creation, or do they just want to absorb the supposed facts? Regardless of the answer, we should be aware that we can receive technical information in various ways from various sources.

2.1 Popular Technical Communication as Ideologically Mediated Discourse

As this chapter shows, technical communication is more than the manuals that come with consumer products or the reports engineers produce. All discourse surrounding technology is *technical communication*:

1) Environmentalists creating literature advocating the benefits of recycling and how to recycle are forms of technical communication;
2) Medical agencies that advocate for public safety through campaigns to wear protective gear engage in technical communication; and
3) Journalists or bloggers extolling the latest and "greatest" consumer technologies reproduce technical communication.

The above types of technical communication rarely are addressed in technical communication classrooms and textbooks, but they are probably the most common way general audiences receive technical information.

The genesis, but certainly not the majority, of most technical information usually comes from expert circles. For instance, the "green movement" promoting sustainable architecture and sustainability in general began with expert-to-expert discourse among scientists and engineers. Various academic journals devoted to environmental engineering and science disseminate knowledge regarding ways to employ energy efficient designs in new technologies and, in the case of science journals, disseminate knowledge on effects of pollution and human encroachment on natural ecosystems. Audiences, both expert and lay, use the findings to promote environmentally sound policies and practices to lay audiences. Obviously, these groups do not expect to communicate effectively to the general public using jargon-laden discourse similar to the discourse in academic peer-reviewed articles. The authors of these lay-audience specific documents report information selectively to adhere to audience expectations and level of understanding.

Another venue for technical communication disseminating knowledge to lay audiences is journalism via "old" and new media. Newspapers, magazines, and broadcasts often report on scientific and technical subjects, but they do not directly communicate expert-level discourse. For example, the nightly news may interview medical professionals about new treatments, concerns, or diseases and excerpt the discussion for the lay audience. On face, this appears to be journalism, which it is, but it is also a form of technical communication. These broadcasts or articles cite (or simply refer to) experts in the field to lend credibility to the information. The journalists or editors may even mention the statistics compiled by an expert, but they most likely would not go into the methods employed to collect the results. The lay audience will accept the report because they feel the experts are credible and do not need or will not listen to a discussion about scientific method or statistical analysis. This type of reporting might be considered "watered down" or "dumb-down" communication, and it is certainly possible to misrepresent data or take facts out of context, but these journalistic reports still communicate technical information and use the rhetoric of technical communication. Invoking scientists, the EPA, engineers, etc. is a strategy for convincing audiences by using credible people.

Scientific journalism, reports about technology, and consumer manuals are typically recalled when one considers the definition of technical communication. Even though those discourse venues are mediated by attention to the needs of general audiences, they are based in scientific/technical epistemology. The next section identifies science fiction as a technical communication genre, which incorporates science and technology as major themes, communicating ideas and values surrounding technology even if the technology is not real (e.g., alien spacecraft).

2.2 Science Fiction as a Source of Technical Communication

In order to embrace science fiction as technical communication, readers must understand that science fiction is more a contemporary cultural comment than a prediction of the future. Predicting the future can be a reality of science fiction (e.g., the submarine in 20,000 *Leagues Under the Sea* [Verne 1870/1995]), but science fiction is a product of its culture. Even if the author (or director in the case of film) intends to predict the future, the author is still a member of a particular culture and absorbs that culture's ideology. Although the genres of science fiction are nuanced, for this discussion, I explain how science fiction acclimates audiences to technology by adhering to a dichotomy. Science fiction narratives often present technology as benevolent or malevolent. The technologies that enslave humanity in films such as the *Terminator* series reflect a disease in a culture that increasingly seems to give up autonomy to machines. In contemporary, real-life factories, robots and other automated technologies are more efficient than human labor and often make certain jobs obsolete, thus, contributing to anxiety about advancements in technology taking away jobs. Also, many science fiction narratives comment on the increased surveillance possible with technology (as is the case in Orwell's [1949] 1984). On the other hand, narratives such as the Star Trek universe of films, television shows, and literature predominantly reflect an attitude of benevolence in technologies. In this universe, technology allows diverse groups (in the form of various alien life forms) to combine forces and explore the universe; medical treatments are expertly diagnosed by waving hand-held devices over patients; and weapons "humanely" stun victims.

Technology as benevolent and always good to pursue is not a universally held value, but it is a dominant value of Western cultures. American culture, especially, holds technological advances highly. Even accursed office products such as fax machines and desktop computers are considered better than the "old way" of communicating through typed correspondence. American media celebrate the release of new technologies by reporting on their initial sale and interviewing early adopters. Contemporary consumer products become characters, celebrities even, for the public much like technology in science fiction narratives can be seen by audiences as characters helping to move along plot elements. Whether the technology creates a setting as in the case of the galaxy-traveling spaceship in *A Hitchhiker's Guides to the Galaxy* series (Adams 1979), which uses an "Improbability Drive," or the robots acting as characters in the many stories by Asimov (1950/2004), technology and science are major *characters* of science fiction. Even though science fiction rarely focuses on possible, at least, contemporarily, technology or science, the narratives still acclimate audiences to perceptions of technology and science and implicitly or explicitly propose that science and technology, which are always advancing, continue to provide solutions and change conditions for humans and society overall. Although believing technology changes culture assumes a technological deterministic paradigm, technology is still socially constructed even if the public believes technology changes life. The commitment to the belief that scientific and technological advancements will improve life is necessary for society to believe in order to allow institutions to pursue advancements.

A major example of a science fiction narrative that explains how humans in industrialized societies commit to technological advancement is Asimov's (1950/ 2004) *I, Robot*. Asimov's novel, which is actually a compilation of short story serials, projects a future where machines take on humanlike qualities in the form of robots. Robots eventually run the world because humans have slowly given over control to the machines, and the robots appear to have united the world. Although Asimov's novel employs some implausible scenarios, the novel speaks to a culturally held attitude—technology will solve problems. The robots act as technological advancements, and the main character, Dr. Susan Calvin, a robopsychologist, narrates the history of robot advancement during an interview with a young journalist. Each chapter is a story about dealing with these robots that do not exactly *behave* as they were programmed. Additionally, each chapter describes a new robot advancement: In the final chapters, super robots create new robots superior to human-made ones. This trajectory of always-advancing technology is a cultural belief, an assumption. Even though the assumption does not communicate technical information, it is a subtle form of technical communication because it acclimates the audience to or reflects attitudes about a socially held assumption—technologies will solve problems in the future. This is discourse surrounding the rhetoric of technology.

Asimov (1950/2004) did not accurately predict the future, but he communicated culturally mediated messages about technological development. The novel mentions labor fears of robots taking over human jobs, but that is a passing theme because each new chapter demonstrates the benefits of new advancements: Labor concerns are never discussed at much length in the novel. The robots and other technologies Asimov "creates" are amplified and atomized technologies of his day. Asimov's vision assumes contemporary technology will advance and become suited for new uses, diffusing to various applications. Therefore, the small nuclear engines Asimov describes are based on the assumption that atomic energy will power future devices with no worry of radiation leaks or, to use a popular media fear, the fissile material falling into the hands of terrorists. Asimov's vision of clean, reliable technology might be naïve, but this vision matches the belief that technology (always) solves problems. Audiences do not question the perspective because as industry and government commitments to technologies show, the public generally supports technological advancement. Science fiction narratives provide visions of what could come and how audiences might deal with technological change. If technological change were not a social value, no technologies or innovations would be pursued.

Certainly, not all science fiction narratives have a positive view of technology, and even *I, Robot* (1950/2004) present human-robot interaction as less than predictable. However, both positive and negative science fiction narratives present technology advancement as a linear, always progressing process. This perspective is important for understanding that science fiction and popular journalistic discourse are forms of technical communication. Both use implicitly or explicitly progress rhetoric to acclimate audiences. All audience members do not have to believe technological solutions will fix contemporary problems, but a large portion adheres to the belief. Otherwise, if the public does not believe in the efficacy of technological advances, society would not have developed apparatus for creating new weapons, communication devices, or medicine. NIH, universities, and various military-industrial-intelligence partnerships are indirectly supported by public desire for advancement of technology. Although science fiction communicates the rhetoric of technological progress, science fiction is a major genre that reflects audience belief that technological advances will solve disease, hunger, and even war. The promise of technologies improving the human condition comes out in *I, Robot* after the company, US Robots, discovers how to build a new type of engine. All they need is a little more time, and the company will perfect "interstellar travel, and humanity has the opportunity for galactic empire" (Asimov 2004, p. 204). In the next chapter readers learn that the engine improved and Dr. Susan Calvin tells the young reporter "now we actually have human colonies on the planets of some of the nearer stars" (Asimov 2004, p. 206). Readers do not get the details of how the company perfected the engine, but, because readers assume technology always improves, Asimov can fiat the engine's construction without worrying the audience will not believe it.

Science fiction does not have to be accurate to acclimate audiences to progress rhetoric associated with technological advancement. Although automation increased from the 1940s, when Asimov (1950/2004) wrote the short stories that became *I, Robot*, and space exploration continues to advance, contemporary robots do not look like Asimov's humanoid vision. However, his work still communicates a technical message about how society perceives technological advancement. As I show in later chapters, Marconi, the popular press, and F. T. Marinetti use science fiction-like narratives in order to demonstrate what potential technology—technology not quite developed—can do by evoking progress rhetoric. Science fiction and general future technology speculation use similar rhetoric. However, it is harder to think of science fiction as technical communication because of the creativity and often fantastical stories expressed through science fiction narratives. Future technology speculation on, for instance, NPR's *Science Friday* series also considers technology possibilities based on contemporary technologies. Because both narratives assume technological progress is inevitable and use progress rhetoric, they communicate a type of technology information—an ideology privileging technological solutions. This ideology is hegemonic and does not need to be held universally within a society.

3 Focus on Non-Canonical Technical Communication

The above types of technical communication are not emphasized in technical communication classes. Contemporary academic bias is towards an employment imperative even if technical communication scholars recognize the rhetoric of technical communication. Technical communication textbooks overwhelmingly stress that students should pursue technical communication instruction because they will need to have career skills. Longo (2000) recognizes this imperative in her review of technical communication textbooks from the second half of the twentieth century (pp. 70–74). I am not arguing these textbooks are wrong—most professions require some kind of technical communication, conveying information about technologies—but workplace imperative overshadows any analysis regarding social aspects of technology. The technical communication field overall does not ignore social constructions of technology, but no popular textbooks cover non-workplace or non-consumer oriented technical communication. Nearly all technical communication textbooks take technology and workplace inculcation as a given—students will occupy technical jobs and communicate through various media. Whether the textbook asks students to mimic prescribed business letter formats (as is the case with most business writing textbooks) or to create specifications for using a technology (as in the case of technical communication textbooks that stress sets of instructions), these textbooks are mediated by a paradigm privileging technology as tools and not as a reflection of culture.

This chapter's main goal is to convince readers to pay attention to non-canonical forms of technical communication. Technical information does not just come from user documentation (e.g., installing a home-entertainment system): Discourse surrounding, including information about, technology is technical communication. General audiences are bombarded by technical communication in overt and subtle ways. For instance, besides the above discussion about condom effectiveness from the NIH report being filtered to a general audience, the public receives technical information from a variety of media—pamphlets, flyers, television, radio, etc.—and those texts, compiled by a technical communicator or communicators, do not always use jargon or complex prose but often cite experts. Appeals of ethos are quite important for scientific and technical information because average readers do not have the time or understanding of complex issues that field-specific experts have; therefore, citing a scientist, using an academic seal, and referring to an industry leader lend credibility to the document's claims. One such pamphlet is a double-sided fact sheet from the National Pesticide Information Center (NPIC) (n.d.) about safety and pesticides for household pets. Besides the NPIC logo, the fact sheet references two credible organizations that establish the document's ethos—the EPA and Oregon State University, where NPIC is located. The fact sheet, using cartoon images of a happy pet and owner, references types of pesticides and poisons inside and outside the home that may harm pets. The document also directly establishes the scientific worth of the information by stating the following: "NPIC is a cooperative agreement between Oregon State University and the U.S. Environmental Protection Agency (U.S. EPA). Data in NPIC documents are from selected authoritative and peer-reviewed literature" (p. 1). This document, aimed at general audiences, filters scientific information from highly technical discourse, peer-reviewed articles, to a two-page, illustrated fact sheet that uses generalized diction to explain how to keep pets away from harmful products. It is technical communication.

Another example of how a general audience receives information about technical subjects through non-technical discourse is when news reporters cover new technologies. The release of the latest mobile communication device causes a media frenzy. Not only do reporters do demonstrations for viewers, they will also interview consumers standing in line to buy the latest gadget (e.g., video game console, camera phone, e-book reader, etc.). These interviews, of course, are possible because of the demand consumers in American culture (but not exclusively America) have for new products: The fact of *planned obsolescence* is not just creating products to wear down after a while; it is also based on convincing consumers to buy the latest, more fashionable product even if their current model still works. Also, the media communicate technical information when interviewing technologists or researchers who observe how practices change when new technologies take hold. In fact, there is an entire Web site devoted to the dangers of texting while driving that uses statistics, legal concerns, images, and even a message from Oprah Winfrey to persuade readers to stop using mobile communication devices while driving (2011 "Texting While Driving"). Although it is inaccurate to claim the entire Web site is technical communication, the Web site references statistics such as, a driver is four times more likely to get into an accident if drunk but eight times more likely to get into an accident while texting (2011 "Texting While Driving Statistics"). Overall, the site uses rhetoric to convince readers not to be distracted when behind the wheel.

Additionally, technical information is communicated by implied or explicit attempts for getting audiences to use common, expected forms of mass communication. For instance, many organizations' Web sites tell audiences to "follow us on…" and provide icons for users to click. The Texting While Driving Web site (2011) even has a link on each page asking readers to "Subscribe via RSS" (Really Simple Syndication) and, ironically, even has a mobile version of their Web site that users can access on their mobile communication devices (presumably not while they are driving). The effect of referencing alternative ways to follow an organization normalizes these communication technologies: This is a subtle form of technical communication, but it is no less of an acclimating type of discourse. When traditional media outlets on television or radio tell audiences to "follow us on" various communication technologies, they are privileging and simultaneously supporting particular technologies they want their audiences to use. Another effect of these communication technologies is that they allow the news to be interactive: Users can communicate with the news outlet. This situation follows the arguments of technology scholars who argued the democratizing potential of the Internet, which was supposed to flatten both access to and production of information (Feenberg 1999; Hawisher and Selfe 1999; Selfe 1999).

The implied messages within forms of technical communication relate to the rhetoric of technology. As Bazerman (1998) defined, the rhetoric of technology "is the rhetoric of all the discourses that surround and embed technology" (p. 387). The cultural-specific messages are part of the discourse surrounding technology. Cultures produce technologies that conform to ideology. One can read technology through semiotics because the technology reproduces and, therefore, signifies concepts a culture understands—even on a subconscious level. In American culture, cars represent freedom, a concept Americans promote. Although an owner's manual does not cover how one attains freedom with purchasing a car, Americans associate cars with freedom. Getting one's driver's license is a rite of passage for American teenagers and provides them with a sense of freedom because they are able to be mobile and drive to school, the movies, the beach, etc. This idea is communicated to teenagers (and others) through popular culture, commercials, and word of mouth. Even though the discourse is not instructive on how to operate a vehicle, as is the case with the owner's manual, the message regarding freedom still acclimates users to technology. It is important to consider this type of technical communication as valid for study because it is potentially a powerful rhetorical aspect of communicating technical information. Another reason why users need to be acclimated to technology is because they are important "actors" or "relevant social groups" that are part of the social construction of technology (Bijker 1995, p. 48). Technology not adhering to cultural values or, at least, not being made to fit cultural values will not become realized.

3.1 Ideological and Historical Discourse is Germane to Technical Communication Studies

The rest of this book demonstrates how technical discourse is socially constructed just as technology itself is—both adhere to cultural values and reflect culture. Marconi's wireless has transatlantic impact before Marconi transmitted wireless signals from Cornwall to Newfoundland. His work created a "buzz" on both sides of the Atlantic. Although Marconi, the popular press, and Marinetti did not just discuss the technical aspects of the wireless, even non-technical discourse about technology is technical communication. Whether discourse comes from experts communicating with one another or journalists, bloggers, and aficionados addressing lay audiences, all discourse surrounding technology is technical communication. The rhetoric used by experts and semi- and non-technical communicators attempts to acclimate audiences to technology. The information is not always accurate, but it still communicates technical information. The discourse surrounding technology is mediated by ideology, so its rhetoric reflects this ideology. Technical communicators are not necessarily conscious of their ideologically saturated discourse. They are members of a culture that, for the most part, believe in technological efficiency.

This analysis is germane to technical communication studies. Historical analyses about technology uncover a time period's values. Many historical technological analyses discuss the technologies' time periods' values and identify discourse surrounding the technology as reflecting ideologies (Bazerman 1999; Bijker 1995; Katz 1992; Sleigh 2007). Historical analyses may illuminate unarticulated assumptions about contemporary technologies. Students and consumers ought to pay attention to these assumptions because even unconscious assumptions mold thinking and value perception. Marconi's wireless has discourse surrounding it that, on face, may seem beyond the scope of traditional technical communication studies. However, that discourse is rhetorical, and technical communication is also rhetorical. My attempt is not to replace instrumentalist-oriented technical communication with rhetorical and historical studies. Instead, this book extends a rhetorical humanistic perspective regarding technical communication.

In order to argue for extending technical communication studies to include more of the above types of analysis, this book must foreground ideas related to the social construction of technology. Readers should consider the next chapter an introduction about how Science and Technology Studies scholars attempt to describe how technology reflects the culture from which it comes. That discussion, which has appeared a few times in the technical communication field, is important for students, scholars, and practitioners in both technical communication and science-engineering fields. Too often audiences assume technology advances because the "best" products win out over inferior ones. However, such a view ignores the social construction and political nature of technology advancement: Technology does not advance on its own; it comes to be based on cultural demands. Communication about technology creation is a component of the social construction of technology and, therefore, germane to technical communication studies even if the communication fails to persuade relevant social groups to adopt the technology. Although technical communication is not inherently objective, most technical communication can be understood as intersubjective, which assumes audiences—both small expert ones and vast general ones—adhere to certain assumptions. These assumptions are socially mediated.

Finally, we return to the beginning of this chapter. The epigraph from Aristotle may be colloquially translated as "preaching to the choir," but the passage resonates with technical communication, specifically how lay audiences receive information. Although maxims have a place in technical communication, the point Aristotle makes that is relevant to the above discussion on technical communication is that listeners (or readers) are predisposed to certain arguments. Nontechnical or lay audiences will not understand highly technical jargon and concepts that have been black boxed by experts. The goal of technical communicators is to present information in general terms. This does not necessarily mean the audience is completely ignorant of the concept, situation, technology, etc. Instead, "people enjoy things said in general terms that they happen to assume ahead of time in a partial way" (Aristotle, trans. 1991, 2.21.15) means the audience has been exposed a priori to discussions of science and technology. As citizens in a technological world, we have assumptions about technologies because we absorb discourse about and have experience with these social artifacts. Audiences do not discover sciences and technologies in a vacuum: Scientists and inventors communicate their work, which gets filtered to lay audiences, but prior assumptions help those experts by conditioning the audience to already accept some information about the specific science or technology (e.g., knowledge of global warming) or a belief in the efficacy of science or technology to solve problems.

To be successful, both technology and discourse surrounding technology have to adhere to audience expectations. One of the most difficult concepts to communicate is that these expectations are ideological and not mechanical/technological. If the origins of technology are demonstrated, this humanist endeavor might convince audiences to pursue socially conscious technologies. Students, for instance, might begin to understand that technology is not a given or that the "best" product does not always come to be. The next chapter discusses the rhetoric of technology and social construction to prepare readers for a specific analysis on Marconi's wireless.

References

Adams, D. (1979). *The hitch-hiker's guide to the galaxy*. New York: Pocket Books.

Appleton, V. (1911). *Tom Swift and his wireless message*. New York: Grosset and Dunlap.

Aristotle, (1991). *On rhetoric: A theory of civic discourse*. New York: Oxford University Press. (G. A. Kennedy, Trans.).

Asimov, I. (2004). *I, robot*. New York: Spectra-Bantam. (Original work published in 1950).

Bazerman, C. (1998). The production of technology and the production of human meaning. *Journal of Business and Technical Communication, 12*(3), 381–387.

Bazerman, C. (1999). *The languages of Edison's light*. Cambridge: MIT Press.

Bijker, W. E. (1995). *Of bicycles, bakelites, and bulbs: Toward a theory of socio technical change*. Cambridge: MIT Press.

Connors, R. J. (1982). The rise of technical writing instruction in America. *Journal of Technical Writing and Communication, 12*(4), 329–352.

Dobrin, D. N. (1985). Is technical writing particularly objective? *College English, 47*(3), 237–251.

Feenberg, A. (1999). *Questioning technology*. London: Routledge.

Feig, C. (20 July 2001). Condom report questions STD protection. CNN.com/Health. Retrieved from http://www.cnn.com/index.html

Hawisher, G., & Selfe, C. (1999). *Global literacies and the World Wide Web*. New York: Routledge.

Johnson, R. R. (1998). Complicating technology: Interdisciplinary method, the burden of comprehension, and the ethical space of the technical communicator. *Technical Communication Quarterly, 7*(1), 210–223.

Katz, S. B. (1992). The ethic of expediency: Classical rhetoric, technology, and the Holocaust. *College English, 54*(3), 255–275.

Kuhn, T. S. (1962). *The structure of scientific revolutions*. Chicago: University of Chicago Press.

Latour, B. (1987). *Science in Action*. Cambridge: Harvard University Press.

Latour, B., & Woolgar, S. (1979). *Laboratory life: The social construction of scientific facts*. Beverly Hills, CA: Sage Publications.

Longo, B. (2000). *Spurious coin: A history of science, management, and technical writing*. Albany: State University of New York Press.

McLuhan, M. (1964). *The medium is the message. Understanding media: The extensions of man* (pp. 7–21). New York: Signet.

Miller, C. R. (1979). A humanistic rationale for technical writing. *College English, 40*(6), 610–617.

Moore, P. (1996). Instrumental discourse is as humanistic as rhetoric. *Journal of Business and Technical Communication, 10*(1), 100–118.

Noble, D. F. (1999). *The religion of technology: The divinity of man and the spirit of invention*. New York: Penguin. (Original work published in 1997).

Orwell, G. (1949). *Nineteen eighty-four*. London: Secker and Warburg.

Rossi, P. (1968). *Francis Bacon: From magic to science*. Chicago: University of Chicago Press. (S. Rabinovitch, Trans.).

Rutter, R. (1997). History, rhetoric, and humanism: Toward a more comprehensive definition of technical communication. *Journal of Technical Writing and Communication, 21*(2), 133–153.

Samuels, M. S. (1985). Technical writing and the recreation of reality. *Journal of Technical Writing and Communication, 15*(1), 3–13.

Selfe, C. (1999). *Technology and literacy in the twenty-first century: The importance of paying attention*. Carbondale: Southern Illinois University Press.

Slack, J., Miller, D., & Doak, J. (1993). The technical communicator as author: Meaning, power, authority. *Journal of Business and Technical Communication, 7*(1), 12–36.

Sleigh, C. (2007). *Six legs better: A cultural history of myrmecology*. Baltimore, MD: Johns Hopkins University Press.

Tebeaux, E. (1996). *The emergence of a tradition: Technical writing in the English renaissance* (pp. 1475–1640). Amityville, NY: Baywood.

Toscano, A. (2011). Using I, robot in the technical writing classroom: Developing a critical technological awareness. *Computers and Composition, 28*(1), 14–27.

Verne, J. (1995). *Twenty thousand leagues under the sea*. New York: Tom Doherty Associates. (Original work published in 1870).

Yeats, D., & Thompson, I. (2010). Mapping technical and professional communication: A summary and survey of academic locations for programs. *Technical Communication Quarterly, 19*(3), 225–261.

Winsor, D. (2002). Ordering work: Blue-collar literacy and the political nature of genre. *Written Communication, 17*(2), 155–184.

Chapter 2
Analyzing Technology to Uncover Social Values, Attitudes, and Practices

> What is called Western or modern civilization by way of contrast with the civilization of the Orient or medieval times is at bottom a civilization that rests upon machinery and science as distinguished from one founded on agricultural or handicraft commerce. It is in reality a technological civilization....If the records of patent offices, the statistics of production, and the reports of laboratories furnish evidence worthy of credence, technological civilization, instead of showing signs of contraction, threatens to overcome and transform the whole globe.
>
> (Beard 1928/1999, p. 97)

> The great nineteenth-century positivists...imagined that the statements of science were going to replace opinions and beliefs about all things....Our century has been the graveyard of positivist ideas of progress.
>
> (Badiou 2005, p. 84)

Technologies are products of the societies from which they come. By reading the technologies a culture creates, we can understand the ideologies of a culture. This chapter examines literature drawn primarily from Science, Technology, and Society studies (STS) that supports rhetorical analyses of technology. Based on these STS studies, I make the three following arguments: First, using the work of theorists such as Charles Bazerman, Bruno Latour, and Wiebe E. Bijker, I argue that technologies are created through a complex system of social interactions where groups affix values to new technological advancements. Second, I argue that rhetorical analyses identify the non-physical ways in which technologies are realized or understood. Third, I demonstrate that social values related to progress have often been affixed to new technological advancements in the twentieth century. In the case of my study of Marconi's wireless, the relevant social groups of the early twentieth century saw the invention as *progress*, modernity's most powerful "god term" according to Weaver (1953, p. 212), and considered the invention an important human advancement. Progress or any other value is not physically built into an invention but rhetorically constructed. These STS arguments enable me to demonstrate the importance of rhetorical analyses of technology in general and Guglielmo Marconi's wireless specifically.

The field of STS includes the philosophy of technology, the history of technology, and the sociology of technology, as well as the rhetoric of technology,

A. A. Toscano, *Marconi's Wireless and the Rhetoric of a New Technology*, SpringerBriefs in Sociology, DOI: 10.1007/978-94-007-3977-2_2, © The Author(s) 2012

which I consider in a later section of this chapter. Obviously, these theoretical frameworks overlap, but I hope to make clear that all are based on examining the socio-political shaping of technology. This work supports my own analysis of Marconi's wireless as a product that held meaning for an audience in a historical moment and was shaped by social interactions. The following descriptions on the frameworks are not meant to be hermetic categories but general ideas.

The philosophy of technology derives from research analyzing the meaning of technology for a culture. Philosophers of technology argue that technology itself appears to be a major context for industrial cultures (Feenberg 1999; Fuller and Collier 2004; Heath and Luff 2000; Melzer 2004; Nye 1994; Rescher 1999; Winner 1986). Melzer (2004) argued that humans are defined by their tool use: *Homo sapiens* are the "the tool-making animal[s]" with "stages of civilization differentiated in terms of the tools men have actually made" (p. 111). Whether one examines the Bronze Age or our contemporary Information Age, one tendency historians of science and technology have is to define human societies by the major technologies they have at their disposal. Therefore, although technology cannot be said to have caused a prevailing cultural attitude, it exists as a defining principle for individuals. No social structure is "free" of technology's influence; therefore, society is governed in some ways by a techno-socio politic. Basically, the philosophy of technology attempts to answer why humans organize themselves, at least in industrial cultures, around technology and the *meanings* associated with technology.

The history of technology aims to define a historical moment by its technology (determinism) or to discover how a historical moment *shaped* technology (constructivism). Some histories of technology provide lists and brief descriptions of technologies from certain historical periods (Bunch 2004; Burns 2005; Cardwell 1995; Glick Livesey and Wallis 2005; Mitcham 1994; Restivo 2005; Rhodes 1999). Other studies focus on a technology in a historical moment or a type of technology from a time period (Cross and Szostak 1995; Downey 2002; Lewis 2004; Löfgren and Willim 2005; Misa 2004; Reynolds and Cutcliffe 1997; Willmore 2002; Yeang 2004). As with any analysis of technology, defining a time period by a technology can be reductive. Not only do historians risk arguing for technological determinism when claiming technology alone shaped an aspect of society, but they also risk creating a "whig" history if their work "presents history as uninterrupted progress, implying that the present state of affairs follows necessarily from the previous" (Bijker 1995, p. 45). Instead of aiming for a grand narrative of technology's influence on history, other scholars look at smaller "revolutions" involving technological change (Cowan 1997; MacKenzie and Wajcman 1999). This book also studies a single technology. The wireless's overall success, the fact that it became the "radio," means it fit well with the social framework of the early twentieth century. Just as the historical context of the early twentieth century is shaped by society, the society accepting a new tool shapes the technology's meaning through cultural attitudes, practices, and values.

The sociology of technology demonstrates how societies—relevant social groups and wider cultural forces—shape and are shaped by technologies.

As mentioned above, placing too much emphasis on the effects of technology on society can risk arguing from an uncritical technological deterministic position. However, a dialectical relationship between society and technology exists, and this relationship is socially constructed. Many scholars analyze specific groups or specific technologies to demonstrate how society shapes what technologies become created based on wider cultural beliefs and attitudes. Additionally, scholars demonstrate how technologies are created socially but on a smaller scale; for instance, many scholars examine the work of a single invention or inventor to demonstrate the social construction of technology (Bazerman 1999; Bijker 1995; Ceruzzi 1999; Johnsom 1995;[1] Latour 1996; Williams 2000). Inventors and relevant social groups must interact within the social framework in order for a technology to be realized. "The social framework," of course, is a product of the culture's attitudes and values that fits or is made to fit social practice.

1 Technology is/as Social Construction

In Latour's (1987, 1988, 1996) work on the sociology of technology and science, he describes how one can view an engineer or scientist's work in situ in order to understand the forces behind technological and scientific "discovery." By visiting laboratories and examining engineering and scientific discourse, Latour and other scholars have uncovered the non-material forces and attitudes that shape what technologies and sciences become realized.[2] Latour's theory of the sociology of technology stems from Michel Foucault's critiques of power in society.[3] Latour (1996) argued that one must consider more than the technological object when analyzing why a technology succeeds or fails; instead, a researcher has to "grab the actors" (p. 89). He wanted "to show technicians that they cannot even conceive of a technological object without taking into account the mass of human beings with all their passions and politics and pitiful calculations"; by doing so "they can become better engineers and better-informed decisionmakers" (p. viii). By applying such a perspective, Latour's methodology lets the individuals' interactions with the technology supply meaning(s) to why a technology fails or succeeds: His "sociology prefers a local history whose framework is defined by the actors and not the investigator" or by grand narratives such as capitalism or transportation theories (p. 19).

Similarly, Bijker's (1995) analysis used a "snowballing" technique to follow actors or "relevant social groups," as he defines them, who interact with a technology in its early stages before it becomes realized (p. 46). Because "[t]echnological development should be viewed as a social process [and] not an autonomous occurrence," Bijker observed "relevant social groups will be carriers of that process" (p. 48). Latour and Bijker's argument that technologies and scientific discoveries[4] are products of social interactions is a foundational principle in STS. This approach assumes technology is not created in a vacuum but constructed through values and practices mediated by the culture in which the technology is created. Because the actors themselves are products of their culture, Marconi would act in ways congruent to the values of industrialized cultures in the early twentieth century. The assumption that socially constructed attitudes may shape technology allows me to view the wireless historically as a product of modernity and larger prevailing attitudes. As (Hiskes and Hiskes 1986) also argue, "social forces and practical goals always determine the current state and direction of technological research" (p. 16).[5] In the case of the wireless, social forces desiring progressive technology and practical goals pushing for communication at sea and to places not accessible by wires drove the wireless's construction. To make the wireless fit larger social values, Marconi and other wireless supporters reconstructed the images of electromagnetic science and technology in relation to the new invention for various audiences.

Such reconstructions are "public statements" that have been carefully crafted and do not immediately count as "the pure world of truth" without an audience's validation (Bazerman 1988, p. 23). Furthermore, as Bazerman argued, "[t]o recognize the rhetorical character of visually transmitted symbolic activity is only to recognize that we live and use our texts in a human world" (p. 23). The rhetoric of the wireless and the actual physical invention are nothing without social interaction. The ideas Marconi and others affixed to the wireless exist within the context of socially maintained ideology. Åkesson (2005) argued that realizing a technology meant more than just being able to produce a working model because technological realization "go[es] beyond the product and depend[s] on relations, feelings, emotions and culturally constructed beliefs about whether something is worth investing in or not" (p. 44). Åkesson's overall argument is that all technologies must be marketed well before they become recognized. Because technologies must fit within a culture's values and attitudes, the "marketing" or PR involved will most likely adhere to or be made to seem to adhere to prevailing cultural beliefs.

I argue the wireless of the early twentieth century was more of a rhetorical reality than a physical one: Pro-wireless discourse portrayed the wireless as "real," but a viable commercial product was not immediately available. Before any technology becomes a black box, the relevant social groups must use, communicate, evaluate, and, ultimately, produce a viable technology. I do not claim technologies solely shape themselves or the practices of relevant social groups; to say that would risk technological determinism. Instead, my study focuses on the wireless's rhetoric as "created" by favorable discourse, rather than on how users physically interacted with the invention. However, I do include the perceptions and, subsequently, the reinterpretations of Marconi and journalists who recount their own experiences witnessing the early wireless's capabilities. The "social interaction" I refer to throughout the book is *discourse*.

Because society ultimately accepts or rejects a technology, how a technology fits or is made to fit into social life depends partly on audience perception. Although I will not analyze early twentieth century audiences' specific responses to technological discourse, I focus on favorable discourse about the wireless because I assume that an ultimately successful technology's positive discourse reveals attributes that help an audience realize its significance. I believe my study on Marconi's wireless helps demonstrate the common rhetorical aspects that "build" a technology to be acceptable to audiences. But before I discuss features of rhetorical analyses, I will briefly describe the dialectical relationship between society and technology from the point of view of relevant social groups. Technologies might not alter the ways people perceive the world or usher in new social conditions, but technologies have altered human practices on small and large scales. For example, the introduction of the automobile helped create a "market" for American suburban life because workers no longer had to live on a "mass transit" line or within walking distance of work. However, would Americans have wanted to live in single-family homes out in the suburbs away from their jobs if they were not members of an individualistic culture with a tendency to expand their living spaces? Likewise, the introduction of mobile phones changed the practice of calling: Now people can talk almost anywhere. Therefore, because suburban life and mobile phones were perceived as valuable, important relevant social groups within the culture approved the new technologies and changed living and telephoning practices to some degree.

2 Relevant Social Groups Affix Meaning to Technologies

As socially constructed artifacts, technologies represent the values of the culture from which they came. For instance, a culture that advocates the "free sharing of ideas" and technical collaboration among its universities and other publicly funded research institutions would most likely pursue technologies to help facilitate such communication.[6] Likewise, a culture at war or simply a military-industrial-intelligence complex perpetually preparing for war will create technologies to improve its defensive or "pre-emptive" capabilities. A culture's needs are often based on the values its people or, more importantly, its institutions hold (Noble 1999; Nye 1994; Rhodes 1999; MacKenzie and Wacjman 1999). Such groups ultimately determine what technologies are created and how they are modified for particular tasks. As Bijker (1995) argued,

> Technology is created by engineers working alone or in groups, marketing people who make the world aware of new products and processes, and consumers who decide to buy or not to buy and who modify what they have bought in directions no engineer has imagined. (pp. 3–4).

No technologies would ever be realized if they were not perceived as conforming to social values and practices.

After Marconi's wireless became a black box—a commercially viable product—the electrical engineering community no longer debated the reality of radio waves, and the radio became a solution to communication problems. Being able to transmit and receive signals, a practice allowed by radio, became the basis for research into radar and navigational technologies (Tarrant 2001, p. 233) and technologies related to broadcasting. But, no matter how well an invention works, a black box technology must also adhere to larger cultural values. Relevant social groups immersed in a particular culture affix meaning to inventions, thus, building technological frames. Bijker (1995) explained that "[a] technological frame is built up when interaction 'around' an artifact begins"; if a frame is not built up in order to "move members of an emerging relevant social group in the same direction," a technology will fail (p. 123). Before users will accept a technology, they must believe the product adheres to social values. These values give meaning to a technological frame. Bijker observed that "[a] technological frame comprises all elements that influence the interactions within relevant social groups and lead to the attribution of meanings to technical artifacts—and thus to constituting technology" (p. 123). Therefore, these frames can be understood as sets of meaning(s) groups affix to technology. From a cultural studies point of view, a frame is a deterministic screen or cultural lens that defines a technology for an individual or a group. In other words, people define a technology's values and uses by the socially constructed heuristic or frame between themselves and the technology that, in turn, helps define a technology.

Although that argument may appear circular—*a technology is defined by a frame that defines the technology*—it actually suggests the dialectical relationships between technology and society. Other scholars define these relationships as "technological regimes" (Nelson and Winter 1982; Rip and Kemp 1998) or "paradigms" (Dosi 1982). According to Rip and Kemp, "[a] *technological regime* is the technology-specific context of a technology which prestructures the kind of problem-solving activities that engineers are likely to do, *a structure that both enables and constrains certain changes*" (p. 340, emphasis added). Although Rip and Kemp focus on engineers, the argument easily transfers to other relevant social groups who are constrained by their own cultural/personal lenses based on how they perceive a technology should be used. Therefore, participatory actions both constrain and construct technological meanings.

Engineers and scientists work under larger social frameworks for producing knowledge. Latour and Woolgar's (1979) study of biologists at the Salk Institute shows the interconnectedness of ideology, science, technology, and history, allowing us the chance to understand broad patterns of cultural beliefs. Salk (1979), in his introduction to Latour and Woolgar's *Laboratory Life*, claimed that "[Latour's] own style of thought was transformed by our [biologists'] concepts and ways of thinking" (p. 13). Salk further commented on the centrifugal and centripetal knowledge diffusion between science and sociology, noting that "[sociologists at the Salk Institute] are coming to recognize that their work is only a subset of our [biologists'] own kind of scientific activity, which in turn is only a subset of life in the process of organization" (p. 13). Latour and Woolgar's goal at the Salk Institute was to uncover "the *social* construction of scientific knowledge in so far as this draws attention to the *process* by which scientists make sense of their observations" (p. 32). Such a process is important to Latour and Woolgar because they argued that sociological studies often examine the scientist without examining the scientific aspects (pp. 23–24). They also argued that researchers ought to pay attention to "'technical' and 'intellectual' terminology," which "is clearly an important feature of [scientists'] activity" (p. 27). This activity comes with a warning against "the uncritical acceptance of the concepts and terminology used by some scientists" because that "has had the effect of enhancing rather than reducing the mystery which surrounds the doing of science" (Latour and Woolgar 1979, p. 29). This process often comes in the form of discourse, such as technical papers or presentations.

No science stands solely on its own merits or the merit of one or a few scientists; instead, science is peer reviewed. Latour and Woolgar used Marvin Harris's terms "*emic* validation" and "*etic* validation" to describe which types of audiences have the final say on scientific representations. According to Harris, *etic* validation derives from "a community of fellow observers"; this group is "the audience who will ultimately assess the validity of a description"; *emic* validation holds that "the ultimate decision about the adequacy of description rests with participants themselves"—the scientists (as cited in Latour and Woolgar 1979, p. 38). The two types of validation help define *when* but not *why* a science or technology exists.[7]

Once validated, a science or technology can be said to exist. However, the existence may only be among a select relevant social group before the "facts" diffuse to other groups. In order to diffuse the facts adequately, a group must persuade others about the value of the science or technology. Such persuasion is rhetorical as Latour and Woolgar (1979) observed. When the scientists at the Salk Institute attempted to make order out of (relative) chaos, they aimed for "the successful persuasion of readers" by stabilizing facts through discourse; the readers, however, "are only convinced when all sources of persuasion seem to have disappeared" (p. 76). According to Latour and Woolgar, observers believed "a systematic, ordered account is attainable" from any observation "no matter how confused or absurd the circumstances and activities of his tribe might appear" (p. 43). Once a fact is stabilized, it appears to have always been there waiting to be discovered (Latour and Woolgar 1979, p. 177). The science seems inevitable in hindsight:

> Once the controversy has settled, reality is taken to be the cause of this settlement; but while controversy is still raging, reality is the consequence of debate, following each twist and turn in the controversy as if it were the shadow of scientific endeavour. (Latour and Woolgar 1979, p. 182)

Audiences accept the science as a reality because "[f]acts are constructed in such a way that, once the controversy settles, they are taken for granted" (Latour and Woolgar 1979, p. 183).

The various mechanisms that bring science and technology to life are not salient features of technical or scientific discourse. One mechanism is granting credit, and scientific credit often goes to whomever "gets there first." According to Latour (1988), "[w]hen we are dealing with scientists, we still admire the great genius and virtue of one man and too rarely suspect the importance of forces that made him great" (p. 14). These "forces" of which Latour speaks are many: economics, philosophy, epistemology, ideology, society, etc. They can also be the groups or teams supporting the "lone" scientist or inventor. The ingenuity and perseverance of a successful scientist cannot be overlooked, but we cannot lose sight of the fact that science and technology are built upon past "discoveries" that engage multiple actors. Also, if one believes a "fact" was just waiting to be found, he or she ignores the construction of scientific ideas. Latour (1988) claimed "[a]n idea… never moves of its own accord. It requires a force to fetch it, seize upon it for its own motives, move it, and often transform it" (p. 16). Scientists often make discoveries based on their fields' histories: "The social context of a science is rarely made up of a context; it is most of the time made up of a *previous science*" (Latour 1988, p. 19). Pasteur and others discovered new science by building on past science, which appeared as part of the social context of a science. Although in popular imagination, lone scientists are seen as revolutionaries, Latour argued that science has to be communicated before the "revolution" can take place (p. 72). Science and engineering need rhetoric to help communicate new discoveries or inventions. Relevant groups use rhetoric to create an image of a technology that reifies the invention through discourse. Relevant groups not only use rhetoric but are persuaded by rhetoric to create meaning for a technology. These groups work to define meaning under the social forces that allow products to take shape within a culture.

3 The Rhetoric of Technology

The study of how technologies are described and realized discursively is known as the *rhetoric of technology*. Bazerman (1998) defined the rhetoric of technology in a rather useful way: "The rhetoric of technology shows how the objects of the built environment become part of our systems of goals, values, and meanings, part of our articulated interests, struggles, and activities" (p. 386). When new technologies become accepted by a population—an acceptance that does not need to be universal—society may change certain behaviors; however, the social forces that propel technologies to be created or simply accepted exist a priori to those technologies. Claiming that a direct causal link exists between technology and social change may be reductive, but we cannot dismiss the power new technologies have on behavior (Feenberg 1999; White 1978; Williams 1990). After all, some people might be persuaded to weave technology into their social practices because they perceive the technology as useful. How a technology is made to seem useful can be understood through a rhetorical analysis of a particular technology's discourse.

Defining the "rhetoric of technology" requires defining *technology*. What makes something a technology? Popularly, *technology* is associated with computers and other "hi-tech" consumer items. Although those items are technologies, they are not the only types of technologies that exist. MacKenzie and Wajcman (1999) claimed

[t]echnology' is derived from the Greek *techne*, meaning art, craft, or skill, and *logos*, meaning word or knowledge. The modern usage of 'technology' to include artifacts as well as knowledge of those artifacts is thus etymologically incorrect but so entrenched that we have chosen not to resist it. (p. 26)

Technologies are thus closely related to new knowledge based on science, techniques, and industry. This knowledge is socially constructed through discourse based on relevant social groups' values and practices. The groups accept and consequently stabilize technologies based in part on the value they perceive in particular tools.

Whether a technology be a physical tool (wrench, hammer, or keyboard) or a mental tool (democracy, management science, or the scientific method), it is often defined as the available knowledge of a civilization closely connected in contemporary times to industry and commerce. The *industrialized* world's economy depends on creating new technologies for growth and "prosperity." These technologies can be managerial or engineering ideas to increase or make production more efficient (management science from Fordism/Taylorism), or they can be actually mass produced products (cars, mobile phones, burgers, etc.) from highly rationalized, efficient systems.[8] Regardless of the type of technology, the idea that technology is some kind of tool related to work or profit predominates. After all, the term "modernization" suggests an entity (such as a nation) acquires or develops technologies that will theoretically improve its economic position.[9] As an important profitable technology for industrial nations of the early twentieth century, the wireless fits the above "scientific/industrial" definition. In fact, many contemporaries claimed the wireless was an important scientific discovery, and Marconi's application furthered national and international industrial goals. These goals were also implicit and explicit demands. For instance, as I demonstrate in Chaps. 3 and 4, British and American imperial goals *demanded* that these nations be in contact with their colonies, and supporters of the wireless discuss this agenda. This is not just an implied need but a military demand, which Marconi and others were willing to fill.

Because society shapes technology, we can locate some social values implicit in representations of the wireless or any technology. Even though values may be affixed to technologies, the rhetoric of a technology often adheres to prevailing cultural values. For instance, many contemporary analyses of technologies often examine how "democracy," a major framework for Western industrial cultures, is affixed to technologies or threatened by new technologies (Feenberg 1999; Montagu and Matson 1983; Selfe 1999; Sikorski 1993; Winner 1986). However, rarely can a technology be said to embody democratic principles of egalitarianism or equal representation. Because the "democracy" label is important to Western industrial culture, relevant social groups often discuss technologies as furthering or not furthering democratic values. Those relevant social groups affix values through socially mediated rhetoric.

The next four subsections discuss how rhetoric helps realize or (re)construct a technology. Specifically, I discuss how Charles Bazerman's study of Edison's light offers a theoretical framework for rhetorically analyzing technology, how technologies are products of systemic forces (i.e., the desire for technologies to be seen as democratic), how technologies become stabilized through rhetoric, and how technologies do not become stabilized.

3.1 Bazerman's Example of Edison's Incandescent Light Bulb

Bazerman's (1999) analysis of inventor Thomas A. Edison's incandescent light bulb studied patents, laboratory notebooks, personal letters, specifications, scientific reports, and popular press articles in order to explain how the light bulb and the Edison System (power stations) were both physically and rhetorically created. Bazerman analyzed discourse(s) surrounding the invention in order to demonstrate the rhetorical acts that led to the technology's creation and acceptance. Edison was a technologist in search of profit and fame, and he, as would Marconi, succeeded in garnering an international celebrity status. Because Edison was a successful (and prolific) inventor, we can assume he invented products that appealed to audiences—potential users. Although all of Edison's inventions were not successful, many were accepted through careful marketing/PR strategies. In other words, Edison and his supporters made his inventions fit the attitudes and values of consumers, causing them to desire his inventions.

Bazerman (1999) traced Edison's growth as a young telegrapher turned industrialist whose legacy continues to this day in the company General Electric. Edison physically and rhetorically "invented" the light bulb, and he kept the public anticipating his invention for economic not technical reasons (Bazerman 1999, p. 181).This marketing strategy suggests that a technology needs more than a physical nature to be realized; a technology's viability is tied closely to its profitability. In fact, Edison had to argue for his invention's future value in order to secure investment (Bazerman 1999, p. 200). He had to show this *value* through fairs, advertisements, demonstrations, letters to stockholders, and interviews with popular press writers, or else the light bulb could have failed commercially as his phonograph did in the 1880s (Bazerman 1999, p. 198). Edison marketed his electric system in places like New York City and Louisville, KY before the light bulb was fully functioning, but the public accepted the "reality" of the technology while the light bulb was mainly a rhetorical construction (Bazerman 1999, p. 219). That is, the public accepted the light bulb before incandescent lighting actually illuminated cityscapes.

Potential users may thus accept the existence of new technical and scientific "discoveries" based on rhetoric. An important rhetorical strategy for technological acceptance is proving value. In fact, most of Bazerman's (1999) work covers the economic strategies Edison used to have the public and investors *realize* that the light bulb existed. Bazerman, arguing from Adam Smith's economic theories, claimed: "[A]ll economic transactions are rhetorical… they are exchanges of value and value is a human discursive construct" (p. 141). New technology does not spring from the earth or from a lab without being conditioned or molded by social forces, and economics is one major force. New technology must work within the current technological system and carve out a niche for itself; it must also "create a dissatisfaction with a current technology" to induce consumers rhetorically to buy new products (Bazerman 1999, p. 142). After all, marketing is "the rhetorical economic work of locating unmet desire and matching potential products to desire" (Bazerman 1999, p. 143). In our postmodern world, desire can be manufactured in order to make consumers want certain products (c.f. Herman and Chomsky 2002). For instance, commercials can portray homeowners as at the mercy of invaders (rhetoric of fear) who are only thwarted by security systems; therefore, security system companies manufacture fear and offer their products as solutions.

Before consumers demand new tools, technology needs to succeed materially and "symbolically (that is adopt significant and stable meanings within germane discourse systems)" (Bazerman 1999, p. 335). Bazerman borrowed the concept of "heterogeneous engineering" from Bijker (1995) to argue that technologies are products of "the coordination and application of many kinds of knowledge and practice, all of which are united and instantiated in the final product" (p. 335). The concept of "heterogeneous symbolic engineering" is imperative for inventors who wish to "[build] enduring meaning and values for the technology they wish to *implant* in our daily lives" (Bazerman 1999, p. 335, emphasis added). Bazerman argued that incandescent light had to "take a place within the discourse and the representational meaning systems of [Edison's] time before it could transform them"; therefore, "the new technology... had first to be built on historical continuities of meaning and value" (p. 350). In the absence of historical continuity and cultural values, technologies do not become realized simply for their own sake. Instead, technologies are often products of systemic forces and rhetorically constructed to be in accordance with prevailing cultural values.

3.2 Technology as a Product of Systemic Forces

Although "technology for the sake of technology" may seem to be the goal of industrial societies, changes in technology "will always be only one factor among others: political, economic, cultural, and so on" (MacKenzie and Wajcman 1999, p. 4). STS scholars and other technology critics engage in broader analyses of technologies by examining how socio-political systems in particular produce new technologies. Technology is a product of systemic forces within and across societies. American citizens, influenced by capitalist ideology, associate technologies with highly profitable companies, but profit is only one factor that causes companies to develop new technologies. This "free market" force helps fuel technological development. Consumers are offered many products, and they often have to replace products as new ones come out. However, consumers are not really free to choose outside of products from dominant companies because those companies together have such a huge market share that consumers would risk being not compatible with the majority of other users. For example, consumers looking to buy computers will most likely gravitate towards Microsoft or Apple operating systems. There is virtually no other competition. Consumers or "the public" are not the only relevant social groups, but technologies do not become realized without them. Often consumers are the recipients to whom more powerful, more invested relevant social groups direct their rhetoric. As with any economic system, the means of production and the means of marketing products are controlled by an oligarchy of *invested* agents working within the techno-structure.

This techno-structure values certain technologies over others. Although large organizations have more influence over technological creation, consumers are not helpless in the face of technological creation; see, for example, human input into such technical designs as "ergonomic" keyboards or "value sensitive design" (Radetsky 2003). However, perception of a technology's value also "conditions" the technologies produced. Winner (1986) argued that, specifically in the United States (but across much of the industrialized world), "[a] fascination with efficiency is a venerable tradition in American life" (p. 46). Americans, Winner went on to argue, have "[a]n eagerness to define important public issues as questions of efficiency" (p. 46). This, in turn, creates a condition where "[d]emonstrating the efficiency of a course of action conveys an aura of scientific truth, social consensus, and compelling moral urgency" (Winner, pp. 46–47). That condition informs Winner's 'Technical Constitution of Society,' which has five "distinctive institutional patterns" excerpted below:

1. "[T]echnologies of transportation and communication... [that] facilitate control over events from a single center....";
2. "[A] tendency for new devices and techniques to increase the most efficient or effective size of organized human associations....";
3. "[R]ational arrangement of sociotechnical systems [that] has tended to produce its own distinctive forms of hierarchal authority....";
4. "[T]he tendency of large, centralized, hierarchically arranged sociotechnical entities to crowd out and eliminate other varieties of human activity....";
5. "[T]he various ways that large sociotechnical organizations exercise power to control the social and political influences that ostensibly control them" (pp. 47–48).

The above five areas constitute some of the systemic forces that perpetuate technological creation. These forces also control users' preferences and organizational behavior.

Because of technology's socially constructed nature and larger institutional control over "hierarchically arranged sociotechnical entities" (Winner 1986, p. 48), a dialectical relationship exists between technology and society. Consumers are not completely helpless users *forced* to buy difficult-to-use technologies or to live under techno-surveilance (as in Orwell's [1949] 1984): A republican-democratic society like the United States can maintain some autonomy. Giddens (1984) explained this phenomenon as *the dialectic of control*: "[A]ll forms of dependence offer some resources whereby those who are subordinate can influence the activities of their superiors" (p. 16). Although Giddens was not specifically describing technology as a "superior," his theory applies because technology must be validated by users or potential users. Unfortunately, this relationship can be skewed in favor of dominant invested parties because an illusion of power to influence actions also exists. According to Giddens, those within the social system are confronted with two "faces" of power: "[T]he capability of actors to enact decisions which they favour" and "the 'mobilization of bias' that is built into institutions" (p. 15). An example of this power situation is bias towards the dominant two-party system in American politics. Citizens can effect change within the government by voting, but the two-party system maintains its dominance, making third parties either nuisances because they often can only take enough votes away from one of the dominant parties or irrelevant because their small size does not allow them to advance their message. Third parties cannot actually expect to win a majority or plurality in presidential elections, but third party candidates have won seats in state and local elections in America as well as seats in Congress. Third parties have tried to win Presidential elections, but no modern ones have come close because the two-party dominance of Democrats and Republicans is overwhelming; therefore, voters carry out their civic duty in accordance to this political bias.

The above analogy of America's two-party system can also be extended to the illusion of "real" democratic choices. Just as voting for Candidate A or B (or sometimes C) constitutes limited choice in American politics,[10] relevant social groups may also position technologies to feign democratic ideals or democratic potential if not viewed critically enough. Democracy, a value (in fact, a technology itself) touted by Western industrial cultures, is often affixed to new technologies. However, "democracy" in the next section is often used synonymously to mean *egalitarian participation*, a reality to which the technologies discussed often do not adhere. Capitalist forces or, simply, the marketplace often decide what technologies become realized because users have already "bought into" the technological system that has conditioned them to continue to buy the latest and greatest products—built-in obsolescence.

What I hope to make clear is that some scholars and critics often *want* new technologies to adhere to democratic principles. They validate a technology's potential by claiming a technology, such as the Internet, supports democratic principles. My purpose is not to debunk the scholars I discuss in the next section but to demonstrate how important the notion of democracy and participation are to rhetorical constructions of technology regardless of whether these terms are accurate descriptions for technologies. Although I could discuss the many popular commercials that advertise the "freedom" of high-speed Internet, I focus on the attitudes of technology critics who promote the Internet as a democratic tool or a potential democratic tool because that analysis relates to rhetoric and the values affixed to hopes for technologies; commercials relate more to consumerist rhetoric.

3.3 Rhetorically Constructing Technology Through Democracy

Technologies are products of "social interaction," but that phrase is misleading if one assumes techno-creation follows an egalitarian inventor-consumer relationship. Although some argue for techno-liberation as a political goal, our current late capitalist, post-industrial information age influences the technologies created. Understanding why "democracy" may help stabilize technologies requires understanding how scholars argue that certain technologies have the potential for increasing democracy. For instance, Feenberg (1999) argued that the Internet is representative of democratization across all technologies. He claimed "computer users in France and the US who introduced human communication on networks originally designed for the distribution of data accomplished a liberating technical innovation" (p. xv).[11] He went on to argue that "[i]n all such democratic interventions, experts end up collaborating with a lay public in transforming technology" (p. xv). Therefore, the onus placed on the "lay public" requires them to converse with the established experts to create a democratic technology. Feenberg's use of "collaborating" is misleading because Internet users are not conversing with those who maintain and promote Internet use.

The Internet's liberation is not congruent to the technological liberation of computer users. Feenberg's (1999) argument that the Internet is a liberating technology does not account for the ways consumers are conditioned to access the Internet through a cycle of dependence on large computer manufacturers. Consumers are conditioned to take part in built-in-obsolescence, and the increasing "user friendliness" of computers further removes users from understanding the structure and programming of complex twenty-first-century networking technology.[12] Users are continually at the mercy of help desks and IT departments when things go wrong or when "important" upgrades must be made. Also, with the exception of virus updates, upgrades done by manufacturers are hardly in the interest of the consumer; for instance, eventually programs bought for one operating system will no longer be supported by future upgrades of the operating system. In the early 1990s and before, when most home computer users had non-networked systems, having an operating system for five or more years was not a serious problem for compatibility. But our current highly networked infrastructure now burdens the consumer to buy new expensive software to harness the capabilities of new hardware. The new software requires new hardware to get online, which further complicates the idea that the Internet is a "democratizing" technology. If true collaboration existed, users would be able to have older systems be more compatible with newer technologies in order to extend a product's use.[13]

Within the field of computers and composition, scholars (most notably, Gail Hawisher and Cynthia Selfe) have embraced a call to techno-democratization. Technology as a democratizing force is reified in "computer lab apparatus": the tools and networks that comprise the totality of computer-assisted instruction—the Internet being the most important tool within the last two decades. Selfe (1999) warned of the "perils of not paying attention" to the increasing knowledge gap between skilled and unskilled students entering college—a gap affecting individuals' future prosperity (p. 4). Selfe's work largely focuses on economic class, but she also covers aspects of the digital divide (the gulf between whites and African Americans online) and the educational backsliding caused by political rhetoric espousing the need to improve children's technical-scientific skills without the necessary public finding. Because these new computer technologies enable students to gain important literacy skills, Selfe argued that the very futures of our students are at stake if they are denied access. Similarly, Henry Louis Gates, Jr. (Gates 1999) argued that African Americans "are failing to gain access to the new tools of literacy," meaning the Internet, and that the Internet is "the most diverse and decentralized electronic medium yet invented" (p. 15). Both Selfe and Gates—reflecting on the Internet circa 1999—underscore the idea that access to contemporary literacy *tools* is a democratic imperative.[14]

On a broader democratic level, Winner (1986) argued that technology, if created in accordance with democratic goals, will support democracy. Winner analyzed undemocratic and horrendously authoritarian technological systems. The often cited example from Winner is that of the architect Robert Moses and his technological Jim Crowism: As "the master builder of roads, parks, bridges, and other public works of the 1920s to the 1970s in New York," Moses "built his overpasses according to specifications that would discourage the presence of buses on his parkways" (p. 23). Such an undemocratic system "limit[ed] access of racial minorities and low-income groups to Jones Beach, Moses' widely acclaimed public park" (Winner 1986, p. 23). Winner argued that Moses' architecture reifies the systemic effects of a racist, classist society; certain groups' access to *public* areas is denied by physical barriers, which "embody a systematic social inequality, a way of *engineering* relationships among people that, after a time, became just another part of the landscape" (p. 23, emphasis added). A similar situation exists in low-income, inner-city dwellings where property values dropped as a result of "urban planning" that sent highways, railways, and other eye sores through predominantly African-American communities.[15]

Racism, a systemic force, helped bring about the technological segregation of Moses's "architecture." Such a situation is representative of how white America treated African Americans throughout history. Moses's setup was simply a microcosm of larger cultural oppression—legally upheld until 1954 but in practice through today.[16] As American history proves, "all men are created equal" is relative to who is in charge. Therefore, "democratic" societies stabilize technology from non-egalitarian social values, attitudes, and practices. Ironically, these undemocratic practices are still in accordance with social values as Winner's example of Robert Moses shows. Also, some technologies "make it" through illegal back channels; for instance, Edison had to deal heavily in the graft of the Gilded Age to see his New York system to fruition; after all, Tammany Hall had to be "convinced" (Bazerman 1999, p. 227). The electric works Edison proposed had to fit certain political agendas. Because electricity is a public works issue, politicians had to support the endeavor, which, in turn, boosts their prestige. Therefore, a working technology is simply one aspect of technological realization. To be successful Edison had "to speak the language of politics... involving patronage, jobs, political support, factional infighting, and perhaps *payoffs*" (Bazerman 1999, p. 228, emphasis added). Such a practice of paying off *civil* servants is hardly an official tenet of democracy.

Besides civilian projects being political in nature, military projects require governmental support, and, as was the case with the U.S. Navy's long-range wireless experiments around 1910, some projects are only possible with state funding. C.-P. Yeang (2004) analyzed how the U.S. Navy "decided to build the world's most powerful radio transmitter in Arlington, Virginia, one that would exemplify American's [sic] military and economic potential" (p. 1). Here is politics of a different sort. Instead of dealing in bribes to carry out radio experiments, the military, which was "in a better position than scientists in university laboratories to conduct long-range radio experiments, for only the state could afford such large-scale projects," had a de facto monopoly on large-scale technological creations (Yeang 2004, p. 3). By 1910, the navy had financial resources because of an internal push towards modernization.[17] They needed a company to outfit ships with wireless technology and construct land-based stations, which meant awarding a contract to the company with the "proper" equipment and contract bid. Politics had a hand in the award. Despite the fact that the National Electric Signaling Company's (NESCO) wireless device "was not in fact able to match the contract's long-distance specification," the navy seems to have wanted "an American over a British or German company"—the British company bidding on the contract was Marconi's (Yeang 2004, p. 9). Because the foreign companies were upset, the navy provided an opportunity for other companies to carry out some experiments (Yeang 2004, p. 9). After all, upsetting possible future companies from helping carry out scientific or technical research would be disadvantageous. The navy was planning more fleet improvements and may have needed commercial support.

The navy did not just decide to outfit ships with wireless because the technology was available; the desire to modernize the navy—a political attitudinal change—occurred because "a number of top-rank officers" believed "[the wireless] could be incorporated into the 'New Navy'" (Yeang 2004, p. 5).[18] The U.S. Navy dealt in favoritism on a micro level but did so because of the appeal to modernization on a macro level. Radios were the high-tech items in the early twentieth century, and they fit the government's idea of what progress meant—modernizing by acquiring new technologies. Having wireless technology onboard meant the navy was being progressive about selecting one of the time period's most important products. What the navy's favoritism and Edison's graft dealings show is that "behind the scenes" forces in which relevant social groups engage also stabilize technologies. Values such as democracy, progress, and modernization are affixed to technologies by relevant social groups, but the reality of such labels does not have to exist. The meanings simply help a technology's favorable perception within a culture.

Relevant social groups affix meaning to technologies that help stabilize them and not stabilize them. Labeling technologies with positive meanings such as progress, democracy, and freedom help technologies; negative meanings such as dangerous, expensive, and inefficient do not help. Many unsuccessful technologies were feasible, but they were not represented to appear practical. Technology studies often focus on the successful products that make it to the market, and the field rarely discusses failed technologies. In fact, Staudenmaier (1985) recognized that the STS journal *Technology and Culture* mainly studied successful technologies (p. 718–719). Since then, two important cases where relevant social groups were not able to affix positive meaning(s) to technologies are Rosalind Williams' study of MIT's *reengineering* (1994–1999) and Bruno Latour's widely popular account of ARAMIS, the failed Parisian rail car system.

3.4 Not Stabilizing Technologies

In one of the few studies on a failed technology, Williams (2000) examines how MIT's reengineering was not accepted by *enough* relevant social groups. These relevant social groups were faculty, staff, administrators, students, and various consultants. The concept of "reengineering was defined as 'the fundamental rethinking and radical redesign of support processes to bring about dramatic improvements in performance" (Williams 2000, p. 643). In order to implement these changes, "a core team analyzed key administrative processes and eventually recommended that eight of them be redesigned" (Williams 2000, p. 643). The biggest, most expensive change "went into installing a new financial system, SAP R/3, which replaced MIT's" accounting system (Williams 2000, p. 634). Williams argued that it was the business side of MIT that forced these changes, and not enough users were happy with the proposed changes. She noted that one colleague "describe[d] staff resistance to reengineering" by stating "'[n]othing is more real than feelings'" (p. 667). Faculty and staff feelings were not "factored in" by the administrators or consultants implementing the changes.

Reengineering failed because relevant social groups did not affix positive meaning to the new technology.[19] MIT's administration apparently was convinced that the new business model of "reengineering," which was popular in the 1990s for increasing productivity (Williams 2000, p. 643), would be embraced by all parties. Unfortunately, the technology, or rather technologies comprising *reengineering*, did not fit within the values and practices of important relevant social groups. Williams explained reengineering's failure as follows:

> While the MIT administration was heavily invested in reengineering, the rank-and-file staff had mixed reactions....Others resented the intrusion of consultants and teams that they considered inexperienced or even incompetent, while feeling that underneath it all reengineering was just about eliminating jobs. The view of the MIT faculty was even more negative....they regarded the whole effort with considerable skepticism and often with outright disdain. (p. 644)

The faculty did not consider reengineering a valuable technology in theory or practice; instead, they "considered reengineering at best a distraction from, and at worst an assault upon, 'real' engineering" (Williams 2000, p. 644).

Williams went deeper into the issue of reengineering's failure by linking it to a greater "problem" of modern technology not being "true" technology. According to Williams (2000), over "the last two centuries... 'technology' has become strongly identified with engineering" (p. 644). However, past engineering was devoted to constructing physical products "cover[ing] a wide range of endeavors" such as "sailing, hunting, weaving, plowing, fighting, cooking, traveling, mining" and so on (Williams 2000, p. 644). She also claimed that "The Massachusetts Institute of Technology trains engineers," and "[f]or conventionally defined engineers at MIT...reengineering is not 'technology' at all but a 'business' or management' application of technology" (p. 644). Because reengineering fit neither the idea of "technology" nor "engineering," it failed to be realized as a technology by enough potential users—most importantly, the MIT faculty. Williams made it quite clear that as a business decision reengineering insulted the "feelings" of those who were supposed to embrace it as a solution for greater productivity. Reengineering simply did not fit the values and practices of the community.

Another technology that did not fit the values and practices of relevant social groups was the almost-completed Parisian commuter rail project, Aramis. The project began in the early 1970s and came to an end in late 1987. Millions of francs were spent on the project, but Matra Transport could not get the system successfully in place. Latour (1996) created situations for a young engineer and a veteran sociologist to investigate what "killed" Aramis. Latour analyzed this failed technology by researching history, economics, behaviors, attitudes, and politics surrounding Aramis. Prototypes were developed, budgets were calculated, and the public was informed, but Aramis failed to become realized. To determine why Aramis did not "work," Latour had the two fictional characters interview Matra employees and members of the Parisian transportation authority and examine press releases, correspondences, specifications, and newspaper articles to construct Aramis rhetorically. These fictional characters "grab actors" in order to produce a sociological account superior to an historical narrative (Latour 1996, p. 89). Latour argued that "[t]he time frame for innovations depends on the geometry of the actors, not on the calendar" (p. 88). Engineers worked on Aramis; time did not. Also, other actors—managers, politicians, accountants—had a part in creating Aramis or, more accurately, not having it realized in Paris. Latour claimed research ought to look to the actors for a more fruitful understanding of the time in which a technology was created: "Grab the actors, and you'll get periodization and temporalization as a bonus" (p. 89). Looking to the actors (or *relevant social groups*) allows researchers to understand when a technology exists. An inventor securing a patent has no more created a "working" technology than a group of engineers with a finished prototype that is not feasible (Latour 1996, pp. 66–67). The fictional characters in *Aramis* travel and interview the key actors in constructing the railcar, asking what factors contributed to any successes or failures at a given moment in Aramis's history.

What Aramis failed to do was become a black box. Aramis did not have an effective support system, and it could not mold human behavior through successful marketing or adapt to behavior through engineers' efforts. One crucial engineering fault was the design that allowed too many individuals the opportunity to choose the direction in which they wanted the cars to go; as M. Gueguen, the Parisian transportation authority director of infrastructures, pointed out to Latour's (1996) fictional sociologist:

People all go in one direction, then the other. If you let people direct their own cars to their destinations, at the end of the day all the cars would be at the end of the line; how would they get back? (p. 90)

On face, this looks to be solely an engineering feasibility issue; however, by considering the fact that the system did not fit the culture's practice, specifically its rush hour practice, this concern is a socially constructed hurdle the ARAMIS system could not address. Gueguen also added that if you tried to fix the above issue "you'd have so many cars, the system would have to be so enormous, that it would cost a fortune" (Latour 1996, p. 90). Also, Aramis's small-car design went against the values of equal access to public transportation: One psychological study on Aramis noted that with the system as proposed "[t]here's no access for the handicapped, or for the blind, or for very tall people, or for luggage" (Latour 1996, p. 187). Additionally, users "expressed fear of being closed in" or "trapped" in a railcar whereas the Metro allowed for greater mobility (pp. 185–186). Again, the potential users' value of security was not a technical but social/personal meaning affixed to what they saw as "confining" transportation technology.

Aramis had a few engineering problems, but not conforming to social values and practices meant it was not realized by the relevant social groups. The Aramis system was a local issue. Although one can say rush hours are similar in many industrial areas, Aramis was specific to Paris. The safety concerns, political "games," and culture cannot be said to be universal. However, industrial cultures do share an almost universal value associating *progress* with technological advancement (Feenberg 1999; Stent 1978; Weaver 1953; Winner 1986). The wireless's stabilization meant progress, and tropes of progress were part of a broader modernist consciousness that encapsulated the "cult of efficiency" promoted not only in technical and scientific discourse but also in avant-garde art.

4 Modernism and Technology

Modernist audiences in the early twentieth century were more likely to be persuaded to accept a technology promoted as a marker of progress. Technologies marked human progress (ion) from the past to the assumed more efficient present and, thus, calling forward to the future. Speed, efficiency, profitability, and movement were all attributes of progress and of modernism. Audiences held the idea of progress in such high esteem that Weaver (1953) argued the term was the industrialized world's quintessential "god term" (p. 212). Weaver defined a god term as "that expression about which all other expressions are ranked as subordinate and serving dominations and powers" (p. 212). Although Weaver analyzed attitudes surrounding progress from a predominantly American point of view, his analysis is not limited to one industrialized nation: Any nation in the twentieth century wanting to increase its industrial power would celebrate progress and the values associated with it. Any rhetorical construction of a technology would be aided by being seen as progressive. The value of *progress* is not in its etymological meaning but in the meaning the modern audience associates with it.

Why would a word that basically means "to move forward" be rhetorically charged to be the god term of the day? That question is easily answered when we see that the era and the mindset of the era affect which words will be god terms. Weaver (1953) argued that a collective stance particular to a time period constructs the god term. Humans define themselves by "[revolving] around some concept of value" or else they "[suffer] an almost intolerable sense of being lost" (Weaver 1953, p. 213). "Progress" was such an entrenched and universal Western goal that anything associated with it made people "socially impelled to accept and even to sacrifice for" the values given to the object (Weaver 1953, p. 214). One would sacrifice "for the 'progress' of the community," and "progress" was "the coordinator of all socially respectable effort" (Weaver 1953 p. 214). Therefore, technologies that espoused progress could arouse nationalist sentiments because the society could point to an object embodying human advancement. New technologies marked a civilization's perceived greatness.

Attitudes towards progress support the "efficiency" practices of Frederick Taylor and Henry Ford, which revolutionized management science and factory production. Although Henry Ford claims not to have had "any Taylor influence over the assembly line," it is impossible not to see Taylor's influence (Beatty 2001, p. 207). John Dos Passos claimed Taylor, who was consumed by efficiency, had "[p]roduction [go] to his head and thrilled his sleepless nerves like liquor or women on a Saturday night" (as cited in Beatty 2001, p. 207). Passos, writing in the 1930s, may be foisting an unwarranted fetishization of Taylor's, but such an attitude would be in accordance with a particular technophile contemporary, Marinetti. Of course, Henry Ford recognized Taylor's ideas or, at least, the importance of efficiency for production.[20] Ford applied Taylor's scientific management to his assembly line and created a new technology indicative of hyper-industrialization at the beginning of the twentieth century.

Because Ford is often credited as *the* inventor of mass production, many audiences uncritically assume he alone is the supreme agent. For instance, Tedlow (2001) claimed Ford alone was responsible for the assembly line (p. 227), but industrial culture already valued efficiency as a goal for technologies. Tedlow ignored the social context by arguing that Ford's investment and not "the market" or "[p]ublic opinion" was solely responsible for the assembly line (p. 227); however, this argument relies too heavily on "a vacuum theory" of technological creation. Social forces must have been "ready" to accept such a streamlined, dehumanizing workplace environment. Tedlow believed Ford "along with Einstein, Freud, Lenin, and a very few others" were "in that class of people who exercised a decisive impact on the history of the twentieth century" (p. 227). What he does not recognize is that these men were products of industrial (izing) cultures. Their genius or impact was congruent to cultural values.

As for Western societies, promoting industrial applications of science was very important. Savvy industrialists and other interested parties founded the Royal Society of Arts and the Royal Institution to advance commercial applications of science. Noble (1999) briefly traced the Royal Society's impact on scientific and industrial promotion, explaining that "[t]here was also a strong connection between the scientific pioneers and early capitalist enterprise" (p. 58). Although Noble's main argument is that these societies' religious convictions extended medieval millenarian beliefs into modernity, he also found the early work of the Royal Society to be "researches focused upon the practical problems of" mechanical and commercial industries (p. 59). These societies promoted technology as humanity's conquest over nature—advancement through techno-evolution. Therefore, technology became a force to be worshipped. It had dominion over the natural world.

Adams (1900/1974) explored the power of machines to become the new spiritual force for humanity in "The Dynamo and the Virgin":

> [T]o Adams the dynamo became a symbol of infinity. As he grew accustomed to the great gallery of machines, he began to feel the forty-foot dynamos as a moral force, much as the early Christians felt the Cross. The planet itself seemed less impressive, in its old-fashioned, deliberate, annual or daily revolution, than this huge wheel, revolving within arm's-length at some vertiginous speed, and barely murmuring—scarcely humming an audible warning to stand a hair's breadth further for respect of power….Before the end, one began to pray to it; inherited instinct taught the natural expression of man before silent infinite force. Among the thousand symbols of ultimate energy, the dynamo was not so human as some, but it was the most expressive. (p. 380)

Interestingly, Adams went on to show his reverence not just for technology, but also for the inventors of new, impressive machines when he claimed, "[h]e wrapped himself in vibrations and rays which were new, and he would have hugged Marconi and Branly had he met them, as he hugged the dynamo" (p. 381). Adams was in complete awe of these new machines, and his account is an apotheosis of their creators. Other modernists glorified technologies as "vehicles" of progress. Marinetti, the founder and leader of Italian Futurism, established speed, efficiency, and ahistoricity as goals for human advancement. Marinetti's manifestos advocate that humans should emulate mechanical characteristics. Although an exaggeration of a social love of progress, Marinetti's works and Adams's enamored state embody the cultural value of progress supported by industrial society.

Because technology was a factor in rapidly changing cultural practices and values, artists echoed or reinterpreted the *meaning* of technology in society. As a contemporary technology, the wireless inspired Marinetti's artistic work. Specifically, he claimed that the wireless influenced his telegraphic style for poetry.[21] However, many artists at the time were experimenting with telegraphic styles; White (1990) suggested "that 'telegraphic' writing was generally 'in the air' in European avant-garde circles" (p. 160). Regardless of how Marinetti was inspired, he reconstructed the wireless through language influenced by this historical modernist moment. Marinetti and the popular press, therefore, reconstructed the wireless for audiences using tropes of progress. Of course, Marinetti's work exaggerates the wireless's possibilities, but those exaggerations show how *progress* shapes his artistic experiments and theories of modernism in general. Part of the Futurist aesthetic was to make bombastic claims about the value of technology and promote its presence as a force of human advancement: According to Marinetti and the popular press, humans advanced or "evolved" through new technologies.

Technology as a marker of human evolution relates to Social Darwinian misreadings of evolution. Evolution is commonly thought to be a progression toward a better species, a higher life form. But Darwin's (1859/2010) theory on evolution does not imply that humans or other creatures become better; instead, they become better *adapted* to their environments. Natural selection is the theory that adaptations enhancing survival cause certain traits to become prominent in a species.[22] Evolution became an important narrative for late nineteenth and early twentieth century Western societies that often feared the opposite—degeneration. As Childs (2000) observed,

> The theory of degeneration threatened Europe with the possibility of a reversion to a less complex and more barbaric form of society. Notions of 'evolution', *'progress'* and 'reform' led to an urgent fascination for their apparent opposites: 'regression', 'atavism' and 'decline'. (p. 39, emphasis added)

Industrial societies had to be seen as advancing through technology; Childs argued, "If a country had not independently achieved an advanced stage of industrialisation, it signified a social and cultural backwardness, an inferiority on behalf of the country's people" (p. 40). Marconi's texts do not speak as narratives against degeneracy, but they do fit the techno-evolutionary narrative that Childs argued came from Darwinian science (p. 36). Marconi and other inventors use *evolution* to mark civilization's progress when they introduce new inventions. The idea that technology signals or is affixed with values of evolution relates to how "democracy" is often affixed *to* or perceived *in* the Internet: Relevant social groups promote evolution and democracy in spite of the reality of such labels. Regardless of the accuracy of certain affixed values, technology still appears to fascinate audiences.

During the Industrial Revolution, new inventions were showcased for popular audiences in World's Fairs. Nye (1994) argued that the World's Fairs embodied the cultural need to move forward because they "served as a site within the transitory present from which the visitor could glimpse the future" (p. 205). World's Fairs showcased progress by showing the public the marvels of civilization; the displays were statements about the importance of technology in society: "They marketed the idea of progress itself, providing an overall impression of coherent historical development" (Nye 1994, p. 205). These popular venues also showcased a nation's technological power—they were expressions of industrial might. And in the political sphere, new movements of the early twentieth century, such as Italian and German fascism, readily embraced hyper-industrialization and the militarization expressed in new technologies. Such inventions as the machine gun, torpedo, dirigible, and even the wireless (which Marconi and the popular press promoted as a necessary wartime tool), as Marinetti shows in his manifestos, would lead Italy into becoming an industrial powerhouse and world military leader. Nationalist sentiments were high in Europe at the time, and technophiles like Marinetti argued a nation's technology established a nation's status.

The desire for modernization led Marinetti to his artistic project: "The liberation of the human body, the liberation of Italian democracy, and the efficient re-working of the machine are one project for Marinetti" (Hewitt 1993, p. 144). Andrew Hewitt argued the systemic reason for Marinetti's *technophilia* "reflects nothing more than the economic underdevelopment of Italy and an understandable fascination with industrialization on the part of the Italian modernists" (p. 146). Marinetti's early Futurist work embodies this fascination and violent patriarchal characteristics of the machine-man.[23] Hewitt observed Marinetti's works as protofascist or indicative of fascist aesthetics, identifying "the figure of the machine" in Futurist art as "the symbol of a specific social and political organization" (p. 146). However, machine- or efficiency-inspired political/social organization is not limited to twentieth-century fascism. The needs of industrialized nations appear to require societies to embrace technological advancement. After all, democratic and socialist nations promoted technology as advancement throughout the twentieth century and through today. From Henry Adams' glorification of dynamos to the contemporary drive in the U. S. for more math and science skills to maintain a world leadership role, technologies and sciences appear to be progress reified.

And the wireless's rhetorical reconstructions show that it embodied the value of progress in the early twentieth century. In order to demonstrate tropes of progress surrounding the wireless, this book shows how Marconi's wireless fit societal values and attitudes supportive of technology by examining a select body of texts espousing *progress* rhetoric. These reconstructions of the wireless through discourse are the first places the wireless exists. Before it becomes a black box, Marconi and others present the wireless as a viable technology in accordance with prevailing cultural attitudes.

References

Adams, H. (1974). The dynamo and the virgin. In E. Samuels (Ed.), *The education of Henry Adams*. Boston: Houghton. (Original work published in 1900).

Åkesson, L. (2005). Trick or treatment: Brokers in biotech. In O. Löfgren & R. Willim (Eds.), *Magic, culture, and the new economy* (pp. 37–45). Oxford: Berg.

Badiou, A. (2005). *Metapolitics*. London: Verso-New Left Books.

Bazerman, C. (1988). *Shaping written knowledge: The genre and activity of the experimental article in science*. Madison: University of Wisconsin Press.

Bazerman, C. (1998). The production of technology and the production of human meaning. *Journal of Business and Technical Communication, 12*(3), 381–387.

Bazerman, C. (1999). *The languages of Edison's light*. Cambridge: MIT Press.

Beard, C. A. (1999). The inevitability of the machine. In R. Rhodes (Ed.), *Visions of technology: A century of vital debate about machines, systems and the human world* (p. 97). New York: Touchstone. (Original work published in 1928).

Beatty, J. (Ed.). (2001). *Colossus: How the corporation changed America*. New York: Broadway Books.

Bijker, W. E. (1995). *Of bicycles, bakelites, and bulbs: Toward a theory of sociotechnical change*. Cambridge: MIT Press.

Bunch, B. (2004). *The history of science and technology: A browser's guide to the great discoveries, inventions, and the people who made them, from the dawn of time to today*. Boston: Houghton Mifflin.

Burns, W. E. (2005). *Science and technology in colonial America*. Westport: Greenwood Press.

Cardwell, D. (1995). *The Norton history of technology*. New York: Norton.

Ceruzzi, P. (1999). Inventing personal computing. In D. MacKenzie & J. Wajcman (Eds.), *The social shaping of technology* (2nd edn., pp. 64–86). Philadelphia: Open University Press.

Childs, P. (2000). *Modernism*. London: Routledge.

Cowan, R. S. (1997). The Industrial Revolution in the home: Household technology and social change in the 20th Century. In T. S. Reynolds & S. H. Cutcliffe (Eds.), *Technology and the West: A historical anthology from technology and culture* (pp. 291–313). Chicago: University of Chicago Press.

Cross, G., & Szostak, R. (1995). *Technology and American society: A history*. Englewood Cliffs: Prentice Hall.

Darwin, C. (2010). *The works of Charles Darwin: The origin of species, 1876 (Vol. 16)*. New York: New York University Press. (Original work published in 1859).

Dosi, G. (1982). Technological paradigms and technological trajectories: A suggested interpretation of the determinants of technical change. *Research Policy, 11*(3), 147–162.

Downey, G. J. (2002). *Telegraph messenger boys: Labor, communication, and technology*. London: Routledge.

Feenberg, A. (1999). *Questioning technology*. London: Routledge.

Fuller, S., & Collier, J. H. (Eds.). (2004). *Philosophy, rhetoric, and the end of knowledge: A new beginning for science and technology studies* (2nd ed.). Mahwah: Lawrence Erlbaum.

Gates, H. L., Jr. (1999, October 31). One internet, two nations. New York Times, p. 15.

Giddens, A. (1984). *The constitution of society: Outline of a theory of structuration*. Berkeley: University of California Press.

Glick, T. F., Livesey, S. J., & Wallis, F. (Eds.). (2005). *Medieval science, technology, and medicine: An encyclopedia*. New York: Routledge.

Heath, C., & Luff, P. (2000). *Technology in action*. Cambridge: Cambridge University Press.

Herman, E. S., & Chomsky, N. (2002). *Manufacturing consent: The political economy of the mass media*. New York: Pantheon.

Hewitt, A. (1993). *Fascist modernism: Aesthetics, politics, and the avant-garde*. Stanford: Stanford University Press.

Hiskes, A. L., & Hiskes, R. P. (1986). *Science, technology, and policy decisions*. Boulder: Westview Press.

Johnsom, J. (1995). Mixing humans and nonhumans together: The sociology of a door-closer. In S. L. Star (Ed.), *Ecologies of knowledge: Work and politics in science and technology* (pp. 257–277). Albany: State University of New York Press. [a.k.a. Bruno Latour].

Latour, B. (1996). *Aramis, or the love of technology*. Cambridge: Harvard University Press. (C. Porter, Trans.).

Latour, B. (1988). *The pasteurization of France*. Cambridge: Harvard University Press. (A. Sheridan & J. Law, Trans).

Latour, B. (1987). *Science in action*. Cambridge: Harvard University Press.

Latour, B., & Woolgar, S. (1979). *Laboratory life: The social construction of scientific facts*. Beverly Hills: Sage Publications.

Lewis, E. E. (2004). *Masterworks of technology: The story of creative engineering, architecture, and design*. Amherst: Prometheus.

Löfgren, O., & Willim, R. (Eds.). (2005). *Magic, culture, and the new economy*. Oxford: Berg.

MacKenzie, D., & Wajcman, J. (Eds.). (1999). *The social shaping of technology* (2nd edn.). Philadelphia: Open University Press.

Melzer, A. M. (2004). The problem with 'the problem of technology'. In D. Tabachnick & T. Koivukoski (Eds.), *Globalization, technology, and philosophy* (pp. 107–141). Albany: State University of New York Press.

Misa, T. J. (2004). *Leonardo to the internet: Technology and culture from the Renaissance to the present*. Baltimore: Johns Hopkins University Press.

Mitcham, C. (1994). *Thinking through technology: The path between engineering and philosophy*. Chicago: University of Chicago Press.

Montagu, A., & Matson, F. (1983). *The dehumanization of man*. New York: McGraw Hill.

Nelson, R., & Winter, S. (1982). *An evolutionary theory of economic change*. Cambridge: Harvard University Press.

Noble, D. F. (1999). *The religion of technology: The divinity of man and the spirit of invention*. New York: Penguin. (Original work published in 1997).

Nye, D. E. (1994). *American technological sublime*. Cambridge: MIT Press.

Radetsky, A. T. (2003, January/February). The perfect piece of toast. *Science and Spirit*, 14–15.

Rescher, N. (1999). *The limits of science*. Pittsburgh: University of Pittsburgh Press. (Original work published in 1984).

Restivo, S. (Ed.). (2005). *Science, technology, and society: An encyclopedia*. Oxford: Oxford University Press.

Reynolds, T. S., & Cutcliffe, S. H. (Eds.). (1997). *Technology and the West: A historical anthology from technology and culture*. Chicago: University of Chicago Press.

Rhodes, R. (Ed.). (1999). *Visions of technology: A century of vital debate about machines, systems and the human world*. New York: Touchstone.

Rip, A., & Kemp, R. (1998). Towards a theory of socio-technical change. In S. Rayner & E. L. Malone (Eds.), *Human choice and climate change* (Vol. 2, pp. 329–401). Columbus: Battelle Press.

Salk, J. (1979). Introduction. In B. Latour & S. Woolgar (Eds.), *(Authors), Laboratory life: The social construction of scientific facts* (pp. 11–14). Beverly Hills: Sage Publications.

Selfe, C. (1999). *Technology and literacy in the twenty-first century: The importance of paying attention*. Carbondale: Southern Illinois University Press.

Sikorski, W. (1993). *Modernity and technology: Harnessing the earth to the slavery of man*. Tuscaloosa: University of Alabama Press.

Stent, G. S. (1978). *Paradoxes of progress*. San Francisco: Freeman.

Staudenmaier, J. M. (1985). What SHOT hath wrought and what SHOT hath not: Reflections on twenty-five years of the history of technology. *Technology and Culture, 25*(4), 707–730.

Tarrant, D. R. (2001). *Marconi's miracle: The wireless bridging of the Atlantic*. St. John's Newfoundland: Flanker Press.

Tedlow, R. S. (2001). Ford vs GM. In J. Beatty (Ed.), *Colossus: How the corporation changed America* (pp. 224–255). New York: Broadway Books. (Original work published in 1996).

Weaver, R. M. (1953). *The ethics of rhetoric*. South Bend: Regnery.

White, J. J. (1990). *Literary futurism: Aspects of the first avant garde*. Oxford: Clarendon Press.

White, Jr. L., (1978). *Medieval technology and social change*. New York: Oxford University Press.

Williams, R. (1990). The technology and the society. In T. Bennett (Ed.), *Popular fiction: Technology, ideology, production, reading* (pp. 9–22). London: Routledge.

Williams, R. (2000). All that is solid melts into air: Historians of technology in the information revolution. *Technology and Culture, 41*(4), 641–668.

Willmore, L. (2002). Government policies toward information and communication technologies: A historical perspective. *Journal of Information Science, 28*(2), 89–96.

Winner, L. (1986). *Do artifacts have politics? The whale and the reactor: A search for limits in an age of high technology* (pp. 19–39). Chicago: University of Chicago Press.

Yeang, C. -P. (2004). Scientific fact or engineering specification? The U.S. Navy's experiments on wireless telegraphy circa 1910. *Technology and Culture, 45*(1), 1–29.

Chapter 3
Marconi's Representations of the Wireless

> *Think of…all the calling which goes on every day from room to room of a house, and then think of that calling extending from pole to pole, not a noisy babble, but a call audible to him who wants to hear, and absolutely silent to all others. It would be like dreamland and ghostland, not the ghostland cultivated by heated imagination, but a real communication from a distance based on true physical laws.*
>
> (McGrath 1902/1999, p. 32)

Ultimately, Marconi's wireless became a successful technology. The relevant social groups in the early twentieth century recognized it as viable and valuable through their negotiations and interactions. Prior to any technology's black-box status, though, is a period in which inventors, potential users, governments, and similar groups receive information about a technology. Although users learn about technologies by interacting with them, discourse is also interaction—it can acclimate audiences to the uses, expenses, and attitudes associated with technology. Normally, we associate technology with hi-tech items, specifically, computers. Before computers became "everywhere," they had to fit into social practices. The computer's success rests partially upon our Information Age's gravitation toward the potentials of databases, search engines, and related hardware and software. For society, computers are not only tools but symbols: Educators and parents believe they allow students greater opportunities to learn; businesses believe they represent efficiency and, therefore, profit; and everyday users, regardless of whether they are frustrated, believe computers are necessities in a world inundated with hi-tech networking. Technologies such as computers do not exist just because they are feasible or operational; they exist because they fit the ideals and practices of a group. Eventually, computers became a necessity to carry out important tasks. Somehow the culture in which they exist accepted them or was made to accept them. Technical communication, the discourse acclimating new users through manuals, articles, and other media, adhering to social values, practices, and attitudes led to users accepting computers.

This chapter demonstrates how Marconi rhetorically constructed his early wireless invention using strategies that show the wireless to be a product of the "built environment" of the early twentieth century. The environment is immediately the British scientific and technical community, but his extended audience stretched across the Atlantic Ocean and to other industrialized countries. Representations of the wireless go beyond merely mechanical or physical descriptions. The wireless was a new technology being presented in a scientific forum before it became the

A. A. Toscano, *Marconi's Wireless and the Rhetoric of a New Technology*, 57
SpringerBriefs in Sociology, DOI: 10.1007/978-94-007-3977-2_3,
© The Author(s) 2012

radio. Marconi's representations constructed an image of what the wireless *might* have become. These presentations were reprinted in technical periodicals and the annals of scientific societies such as the Royal Society of Arts, London, the Royal Institution of Great Britain, and the Smithsonian Institution. These texts that contain Marconi's presentations show the wireless as a new product signaling human evolution, creating an economic enterprise, supporting the military, and increasing the efficiency and scope of (mass) communication. The wireless "fit" the social framework of the early twentieth century because it was an advancement presented as a new communication device that would allow humans another form of control over nature. Marconi promoted the wireless as a tool to make the seas safer for travel, insure cargo better, and connect ocean goers with "land news."

Marconi's representations of the wireless demonstrate how rhetoric builds meaning into technology. Selecting technical presentations, which later became technical documents and archived texts, is an appropriate way to discover the rhetoric of technology. Marconi presents his wireless in these texts as an ideal product for society and an invention indicative of modern progress. The texts, however, are but one of several types of rhetorics for the wireless. In the next chapter, I will examine how the popular press uses another rhetoric to (re)present the wireless for a popular audience. Additionally, patent documents and lab notes are rhetorics because they represent the wireless symbolically—through text and images. My study notes those symbolic rhetorics but focuses on a specific type of discourse aimed at a wider audience. Although much can be said about the forums in which Marconi presented his findings, this chapter concentrates on the texts reprinted from his presentations. I do not discuss all of Marconi's presentations on the wireless throughout his career. The focus of this chapter is on the early years just before the wireless became viable both commercially and culturally. The goal is not to claim Marconi's language led people to adopt this wireless. His agency is discussed in much greater detail by others (Corazza 1998; Garrat 1994; Hancock 1950/1974; Hong 2001; Marconi 1982; Marconi 2001; Tarrant 2001). Instead, my goal is to sample how his language identified the wireless as a product of the early twentieth century. We can never know for sure exactly how all audiences received information about the wireless. These technical presentations were not aimed at the masses as journalism was. However, the presentations are part of the discourse surrounding the wireless, and my analysis presents a close reading of a few texts. No communication act is devoid of rhetoric, so all discourse surrounding the wireless, including Marconi's presentations, builds the wireless's meaning—its rhetoric. These presentations reveal a snapshot in time of the rhetoric of the wireless and suggest how discourse defined invention.

In order to demonstrate how the wireless fit within an industrial culture that favored "progress," I have selected three of Marconi's early presentations to examine how he discusses or represents the wireless as a viable early twentieth-century technology through discourse in the following reprinted presentations: (1) to the Royal Institute of Great Britain on February 2, 1900 ("Wireless Telegraphy"); (2) to the Royal Society of Arts, London on May 15, 1901 ("Syntonic Wireless Telegraphy");[1] and (3) to the Royal Institute of Great Britain on March 3,

1905 ("Recent Advances in Wireless Telegraphy"). The three presentations are instances where Marconi rhetorically represents the significance of the wireless. He delivered them to an important community and spoke to concerns outside of science and engineering prior to the wireless becoming a literal and figurative black box—the radio. Marconi delivered the presentations before highly technical/ scientific audiences, and his discourse conveyed more than technical information during this early stage of the wireless. Marconi presented much of the potential of his new system along with current successes: He discussed how ships communicated with other ships and land-based stations; he mentioned how people and cargo were saved by distress calls; and he prophesized that the wireless would change how information was disseminated. The forums where he delivered his presentations were important—the Royal Institution and the Royal Society of Arts, London was products of a culture invested in scientific and technical progress; they were clearinghouses and think tanks for "mechanical arts." Their goal was to promote the practical application of science to human endeavors. These groups and groups similar to them—Birmingham's Lunar Society, the French Societe d' Encouragement pour l'Industrie Nationale, and the American Society for the Promotion of the Useful Arts—were technical venues where scientists and engineers displayed their work and promoted ingenuity, and they were also founded by inventors, industrialists, and businessmen with entrepreneurial-industrial pursuits (Noble 1999, pp. 77–79).

These three presentations are not meant to be definitive examples of the rhetoric of the wireless; instead, they represent Marconi's engagement in the rhetorical construction of a technology prior to its becoming a black box. Although Marconi was a widely known and respected inventor in Europe and America, between 1900 and 1905, he did not have a widely successful commercial product. He did cross the Atlantic in 1901 and had ships outfitted with the wireless, but he was still attempting to prove the wireless was beneficial (and real) in these early years. Also, his presentations supported his ownership of the wireless, which I will demonstrate later in this chapter as important to establishing himself as the future "father of the radio." Marconi did not just speak about the technical details of his invention; instead, his presentations show the wireless as a technology embedded in the values of a culture. Specifically, the texts reveal four topoi related to Marconi's rhetoric: cultural pride associated with technological advancement, expectations and current successes of the wireless, the economic viability of the wireless, and Marconi's ownership of the wireless. Additionally, I do not analyze these presentations to determine the context of the modernist condition. Although these presentations are mediated by modernist values associated with technology and progress, I describe the time period further in Chap. 5. This chapter's focus is on Marconi's language to technical and semi-technical audiences. Whether or not Marconi excited all audience members, thus leading them to accept his genius, is irrelevant. His 1909 Nobel Prize in Physics denotes his historical importance. These presentations are instances of discourse surrounding the wireless, which help define its rhetoric.

The rest of this chapter demonstrates how Marconi fit his wireless into the technical discussion and how the four topoi above, combined with other rhetorical strategies beyond physical representations, define the rhetoric of the wireless. First, I show how the wireless fit into the early twentieth century's scientific-technical framework and the importance of Marconi's texts and presentations for the wireless's rhetorical construction. Second, I examine how Marconi builds a scientific ethos to attempt to establish his credibility. Finally, I demonstrate how the above topoi create an image of the wireless.

1 The Wireless's Place in Early Twentieth-Century Science and Technology

Marconi's presentations to the technical and scientific community represent the wireless as *human* advancement. Western industrializing societies were ready to welcome technology for the sake of newness, and Marconi's audiences gathered in these technical/scientific forums to learn about what was new in the electromagnetic arts. His using wireless signals to cross the Atlantic Ocean made him a celebrity, but he had rivals and critics who did not feel he had a good working technology (c.f., Weightman 2003). The forums, therefore, were necessary for gathering support for the wireless's existence. Audiences had to believe such a device—one transmitting and receiving *invisible* waves—was possible. Marconi had investors and a board of directors in early 1900 (Hancock 1950/1974, p. 25), so he had investors and clear economic goals. Therefore, the presentations were most likely conversations solidifying the wireless's scientific merit. The scientific merit of the wireless and its economic practicality were not mutually exclusive.

Marconi's mentioning of the economic possibilities of the wireless in his presentations shows that the application of science to practical endeavors reflects how his rhetorical choices adhere to certain values of industrialized society—economics. Marconi might not have been consciously constructing his presentations to "fit" traditional scientific discourse, but his presentations also support modernist, industrial, Western values for technology. Furthermore, the texts convey the wireless's rhetoric because they discuss the wireless's potential nature. Marconi, the technologist, does quite a bit of marketing in these texts, which reinforces the fact that his words were not simply to discuss the physical makeup of his wireless, but also its economic potential.

Marconi demonstrated the wireless's value as a viable product to enhance contemporary communication, positioning the wireless as a better communication tool than cables. Such positioning is really a public relations campaign, and Marconi staged some important wireless transmissions to intrigue the public: He had Queen Victoria send a message to her son, and, a few years later, he had Theodore Roosevelt and the King of England send messages to each other; both events were covered in the popular press. He commented on his successes at sea reporting on yacht races during the presentations; of course, this has no scientific merit but is certainly exciting. As Edison did before him with electrical works (Bazerman 1999), Marconi was looking to create a market for his system before his system became a reality. Marconi's early system (circa 1900) needs to be distinguished from the future radio: The radio is an artifact and a black box technology; the wireless, is an idea comprised of a transmitter, receiver, and various other parts that connects people and stations without wires. The early wireless is a scientific-technical-business product: "Marconi's work-network nods respectfully to the scholarly achievements of scientific discovery, conceived as properly application-less, but simultaneously pirates its apparatus, adapting it to a precedent technology pregnant with extant applications, telegraphy with wires" (Jaffe 2009, p. 19). Therefore, Marconi's application becomes positioned to be the next advancement over cables and wires. The texts where the wireless "exists" in the early twentieth century are artifacts on wireless negotiations between Marconi and the technical community. They reveal more than the invention's mechanics; they reveal potential economic viability.

The need to discuss a technology's economic viability is made clear by Charles Bazerman, who noted "technology must always overtly adhere to the marketplace, political ambitions, and personal desire" (Bazerman 1998, p. 383). Those who credit Marconi with inventing the wireless attach significance to his economic vision for the wireless. Thus, Judge William K. Townsend—supporting Marconi from a necessary legal position—agreed that Marconi was a proprietor of wireless knowledge because he "appreciate[ed] that this new current was destined to carry onward the freight and traffic of world commerce" (as cited in Hancock 1950/1974, p. 6). In this particular case, the court granted Marconi patent rights because he took the science of hertzian waves (also called "wireless signals" and, later, "radio waves") and created a *commercially* viable system for harnessing their power. Marconi's presentations go beyond patent rights. He does use patent rights to claim ownership in his presentations, but he also shows that he is worthy of making scientific claims and constructing a reliable product. Patents provide credibility in legal terms, but Marconi's rhetoric attempts to establish him as an important figure scientifically.

2 Building a Scientific/Technical Ethos

Marconi (1900) presented the wireless as a new incarnation of electromagnetic science, a field dating back nearly a century from his early work. He demonstrated that his work was built upon an old field in his introduction to his 1900 presentation "Wireless Telegraphy," where he brings up six major scientists: Andre-Marie Ampere, Joseph Henry, Michael Faraday, James Clerk Maxwell, Heinrich Hertz, and Sir John Ambrose Fleming (p. 287). Two of the six are familiar to most audiences, but all are familiar to technical audiences. Ampere was a French physicist of the early nineteenth century for whom the ampere (amp) was named. Heinrich Hertz was a German physicist who first discovered what we know call radio waves. His name today is mainly associated with measuring cycles per second (Hz). Marconi's references to past inventors show him standing on the shoulders of greatness. In part, he is borrowing a scientific ethos by showing he understands the field's rich history, but he is also situating himself in this history. This strategy attempted to solidify his credibility while promoting the wireless as an important new technology.

Of course, Marconi's accolades to past scientists could just be out of respect, but they also introduce the audience to a technology with an important history. The introduction to Marconi's (1900) presentation offers the following descriptions about three important scientists in the field: James Clerk Maxwell's "splendid dynamical theory of the electro-magnetic field"; Heinrich Hertz to whom "[w]e can not pay too high a tribute to [his] genius"; and Professor Hughes "the eminent electrician" (p. 287). Marconi also uses Professor Fleming's words to praise Hertz when he claims Hertz's discovery is the "greatest insight into the hidden mechanisms of nature which has yet been made by the intellect of man" (1900, p. 287).[2] Not only does he simultaneously praise Hertz and invoke Fleming, but Marconi adds a mystical element to his own work: Marconi had now harnessed the power of "hidden mechanisms of nature"—sending wireless signals. Marconi's language seems to suggest a way of comprehending these invisible waves, which are part of the built environment, part of technological advancement. Later, the popular press would describe Marconi as a wizard (like Edison and Baekeland) whose efforts were as magical because he could conjure up the forces of nature. In his presentation here, it is nature that he controlled: "A century of progress such as this has made wireless telegraphy possible. Its basic principles are established in the very nature of electricity itself. Its evolution has placed another great force of nature at our disposal" (1900, p. 287). Marconi's words "a century of progress" referred to the science of electromagnetism, but, rhetorically, those words connected the wireless to human evolution. After all, *nature* is at the disposal of humanity; therefore, using wireless telegraphy marks an evolved species.

The scientists Marconi invokes are the "geniuses" who harness nature's power, and Marconi implicated himself in this important work. In another presentation, Marconi showed that he is continuing in the rich tradition of eminent scientists when he mentioned "the memorable researches and discoveries of Faraday carried out in the Royal Institution" (1905, p. 131). Because Michael Faraday was the founder of these weekly meetings at the Royal Institution, Marconi's invocation works in two ways: (1) he implicitly linked himself with Faraday—the man who first worked in electromagnetism, and (2) he positioned his and Faraday's work in electric communication as an important field—a foundational field—for the Royal Institution. He finished his brief historical prelude by mentioning Maxwell's "wonderful mathematical theory of electricity and magnetism," which was "inspired by Faraday's work," and explained in the text how, in turn, Maxwell's work influenced Hertz, who "in 1887 furnished his great practical proof of the existence of these *true* electro-magnetic waves" (1905, pp. 131–132, emphasis added). Marconi described each of the above scientists' works as "memorable," "wonderful," and "great," respectively. In the context of scientific progress these *memorable*, *wonderful*, and *great* accomplishments are gifts to humanity.

By adding his name to this rich history, Marconi positions himself as an important figure for technological progress: "Building on the foundations prepared by these great men, [I] carried out in 1895 and 1896 [my] first tests, with apparatus which embodied the principle on which long-distance wireless telegraphy is successfully worked at the present day" (1905, p. 132). Marconi built himself into the history of the wireless again in this presentation by claiming that "a new method adopted by the author in 1898...was a step in the right direction" for improving syntonic wireless telegraphy (1905, p. 133). In this particular instance, Marconi had stations in close proximity to carry out syntonic wireless telegraphy experiments. This alone signaled he was engaged in science, but, in a stilted instance of technical communication prose, Marconi further established his scientific ethos by claiming with these stations "a very important and effectual limit to the practical utilization of wireless telegraphy would be imposed" (1905, p. 133). He then explained in technical terms how he was able to correct his system so that transmitters and receivers could be tuned correctly, thus allowing messages to be sent to the correct place without being lost, jumbled, or re-routed (1905, p. 133). Therefore, his particular methodology led him to make a scientific discovery.

Marconi built his scientific ethos in order to have credibility with the scientific and technical communities. His audience was more inclined to accept his invention's value and practicality (and even existence) if he adhered to the discourse community's "rules" for presenting new knowledge. And presenting a new invention as a viable technology, one not ubiquitous as the radio is today, required Marconi to present himself as scientifically credible. Wireless signals or, as they were called, "hertzian waves," were not fully understood in the early twentieth century. Part of Marconi's discourse was dispelling myths about the limits of hertzian waves by countering with scientific "facts." Building upon the work of those before him contributed to his establishing a scientific persona and allowed him to be seen as an authority who will accurately portray the wireless; he needed to become the wireless's mouthpiece, and a believable mouthpiece would have to be scientifically knowledgeable. His rhetoric further constructs the wireless not as a trivial, dubious gadget, but as a practical apparatus fulfilling a perceived social need ending a century of scientific progress.

Marconi also built his scientific ethos by adhering to a few "standard" scientific-paper practices. While Marconi attempted to create a consensus about the science of wireless signals, he did not downplay his involvement in his research except through his use of passive voice and third person speech (which may only be an editorial stance for the journal in which it is reproduced). Still, his narrative borrows the scientific ethos Gilbert and Mulkay (1984) described in scientific papers where conclusions (or accomplishments for Marconi) seem to follow "unproblematically from empirical evidence" (p. 46). Marconi spent little time on the problems or failures he encountered when constructing the wireless. Instead, he discussed the successes as if they were simply waiting to be discovered. According to Aristotle (trans. 1991), this rhetorical strategy is a necessary component of persuading an audience:

[There is persuasion] through character whenever the speech is spoken in such a way as to make the speaker worthy of credence; for we believe fair-minded people to a greater extent and more quickly [than we do others] on all subjects in general and completely so in cases where there is not exact knowledge but room for doubt. (1.2.4)

Marconi's strategy was to portray both himself and the wireless as credible. The technology had not yet been universally recognized by the technical community, so Marconi built a scientific ethos while rhetorically constructing the wireless, an emerging technology without "exact knowledge" but much doubt.

The scientific aspects of his presentations were necessary because he was convincing an audience—an interested but skeptical one—that wireless signals did, in fact, exist. Once the technology becomes a regular feature of an established system or tool, the technology does not need to justify its existence scientifically. For instance, the internal combustion engine "runs" on the idea that gasoline can explode in a controlled way in order to drive pistons and other mechanisms to make an automobile move. The science of such a tool had to be hypothesized before the practical application—such as a diesel engine—could be created. The wireless followed a similar timeline. Marconi hypothesized that he could send wireless signals, and his experiments enabled him to create the wireless. But, before the final application existed, Marconi promoted his "flawless" hertzian wave experiments through scientific rhetoric.

All of Marconi's tests were not successful, but that fact is nearly impossible to extrapolate from his presentations. Even the historical accounts seem to gloss over his failures and present events as if they happened more or less flawlessly (Corazza 1998; Hancock 1950/1974; Tarrant 2001). Marconi, as would any inventor, had to "try out" different ways to transmit and receive wireless signals. Although some accounts mention Marconi having trouble with a test or tests, the accounts normally claim factors other than the technology were to blame (c.f., Hancock 1950/1974; Tarrant 2001). In the 1900 presentation, for instance, Marconi blames human error when explaining why his system was defective during the Boer War in South Africa. Some assistants went to help, "[b]ut when they arrived at De Aar, they found that no arrangements had been made to supply poles, kites, or balloons" (Marconi 1900, p. 295). The crew had to use inferior makeshift kites; therefore, "[t]he results which [Marconi's assistant] obtained were not at first altogether satisfactory, but this is accounted for by the fact that the working was attempted without poles or proper kites" (Marconi 1900, p. 295). Additionally, Marconi made it clear that the wireless was still a good system:

It is therefore manifest that [the wireless operators'] partial failure was due to the lack of proper preparation on the part of the local military authorities, and has no bearing on the practicability and utility of the system when carried out under normal conditions. (Marconi 1900, p. 295).

The problems above could have been harmful to Marconi's reputation (ethos) and, therefore, harmful to the wireless system. Imagine the public hearing that Marconi's assistants using any apparatus failed to establish communication. Regardless of inferior equipment, observers could still perceive the results as a system failure, and that perception would be hard to overcome. Marconi (1900) himself stated "that if I had been on the spot myself I should have refused to open any station until the officers had provided the means for elevating the wire, which, as you know, is essential to success" (Marconi 1900, p. 296). Marconi creates the idea of a flawless invention to convince his audience of its potential existence. Of course, having a viable military application would bolster Marconi's economic potential, so he offers with confidence "that before the campaign is ended wireless telegraphy will have proved its utility in actual warfare" (1900, p. 296). Marconi even foreshadows the wireless's military usefulness when he invites the audience to "agree with me that it is much to be regretted that the system could not be got into these towns prior to the commencement of hostilities" (1900, p. 296).

Marconi was positioning the wireless and himself favorably. The scientific ethos he built was an enhancement to his rhetorical construction of the wireless. But being credible was not enough to have the wireless realized. Marconi had to show how the wireless fit with the social attitudes of his audience. Marconi spoke to the audience's desire for advancement, another cultural trait of Western society.

3 Cultural Pride Through Technological Advancement

Technological progress is a major trope for modernity. Inventions were seen as progress, and *progress* was an important cultural attribute to associate with products of modernity. Marconi himself presented his invention as a monumental technology, one that would define the future of communication. By claiming that his work with the wireless had enormous, even unthinkable, potential, Marconi adhered to the belief that technological capabilities were beyond comprehension. He also reaffirmed the popular belief that technological advancement marks human progress. Human evolution appears to be implicit in positive representations of technologies. The wireless's feats became human feats. Marconi portrayed the wireless as a "record breaker" to show where his work fit in the history of electromagnetic science and where the wireless would take humanity in the future.

To many in the early twentieth century (and today), technological progress had *virtually* no limits, and that realization meant celebration for a culture. Nye (1994) mentioned how Americans celebrated technological feats, viewing them as sublime. The opening of the Erie Canal and the start of construction on the Baltimore and Ohio Railroad were both firsts that Americans celebrated, and they did so on the Fourth of July (Nye (1994), p. 47). Nye goes on to say "the canal was understood as a product of democracy," meaning the culture shared in the glory of having contributed to its creation (p. 36). The canal stood as a testament to American progress. Technologies were a part of the national consciousness and new displays were a chance to celebrate human endeavors. Being the first to accomplish or create something is significant; for instance, a person could cross the English Channel by swimming, a pilot could cross the Atlantic Ocean by flying an airplane, or a newspaper could be published in a place where it never before existed. Marconi accomplished all three of the above events using wireless signals, and all were major technological feats.

Marconi's texts emphasize the *firsts* in wireless communication, which signal additional steps in human advancement. Technological firsts are important to a culture because they are instances of cultural pride and exciting stories for techno-societies. The "first" Marconi mentioned in his 1900 presentation was a relatively small one compared to walking on the moon or even sending the first wireless transmission across the Atlantic, but being able to print "a small paper called the *Transatlantic Times*" was a major commercial step forward for his system (Marconi 1900, p. 294). This event revolutionized communication during travel, and Marconi took pride in announcing that "*the first* instance of the passengers of a steamer receiving news while several miles from land...seems to point to a not far distant prospect of passengers maintaining direct and regular communication with the land" (1900, p. 294, emphasis added). Again, Marconi's discourse directed the audience to the future, which was an important rhetorical move for a culture inundated with new technological advancements because the cultural expectation was that technology will always improve and, therefore, improve life.

Although firsts can bolster cultural pride, they may also ignite intercultural dispute: When invited to celebrate the wireless's fiftieth anniversary, the Russian ambassador to Rome declined because the Soviet Union had already celebrated the wireless's creation two years earlier by honoring the man they believed to be the true inventor of the wireless—the Russian Alexander Stepanovitch Popoff (Hancock 1950/1974, p. 2). H. E. Hancock refutes that Popoff was the "father of the wireless," claiming "Russian psychology" is to blame for the ambassador rejecting Marconi's historical significance as the wireless's inventor; but the cultural capital of being the first is quite apparent (p. 2). Popoff had technically been the first to send wireless signals, but Marconi is credited with being the first to put the wireless into commercial use, which seems to give him his stature among many biographers (Bucci et al. 2003; Corazza 1998; Crowther 1954; Garratt 1994/2006; Hancock 1950/1974; Hong 2001; Jensen 2000; Kraeuter 1990; Marconi 1982, 2001; Tarrant 2001). The commercial first of Marconi's wireless allowed him "a communicative application for credit-taking" (Jaffe 2009, p. 19). Here is where the technologist "wins" over the scientist according to Aaron Jaffe:

> To say that Marconi—or, for that matter, Hertz, Tesla, Popov, Branly, Bose, Lodge, Fleming, Fessenden, DeForest, or Armstrong—invented radio is not to say he did so *ex nihilo* but instead that his name became 'black-boxed' with a certain modification of the application's component configuration (p. 19).

Additionally, Marconi's black-box status or "brand" has support from audiences that associate the wireless with him.

Being the first to envision the economic potential of the wireless, or, at least, being the first recognized by history, bolstered Marconi's credibility. When he discussed his "firsts," he furthered his scientific ethos, allowing his rhetorical construction of the wireless—as a monumental technology—to have more credibility. Here the distinction between the wireless and the radio is most crucial: The radio is a technology that has not been in question in nearly a century; the wireless in its pre-black-box stage needed to be accepted by the technical and scientific communities. Showing his expertise to the electromagnetic community was part of Marconi's rhetoric, and he struggled not only with showing his new invention in a positive light, but also with debunking established scientific "truth." In fact, even Edison, an established techno-celebrity, did not believe Marconi crossed the Atlantic because such a crossing was contrary to the accepted scientific belief that radio waves could not pass through the curvature of the earth (Tarrant 2001, p. 63). Eventually, Edison and the rest of the scientific community revised their earlier doubts and accepted Marconi's triumph. Therefore, Marconi building his scientific ethos was a necessary aspect of his rhetorical maneuvering—speaking as an expert who had to convince audiences that he had a new product that worked and could be used for profitable, practical applications. Later in this chapter I will show how Marconi directly dispels this myth, but it is important to emphasize that his accomplishments (e.g., crossing the Atlantic) did not stand by themselves. Marconi had to convince the audience he accomplished this transatlantic feat. After all, how does one convince an audience to believe in *invisible* waves?

Another way Marconi built a scientific ethos was by detailing his new improvements to emphasize his expertise and success. Marconi described a "magnetic receiver" that increased the distance of wireless transmissions, thus reducing doubt about his wireless claims. In his 1905 presentation, he told the audience about this improvement now being "used on the ships of the Royal Navy and on all trans-Atlantic liners which are carrying on a long-distance news service," and "[i]t has also been used to a large extent in the tests across the Atlantic Ocean" (pp. 137–38). In other words, Marconi's improvement was already in practical use. He also reaffirmed his status within the scientific community by again mentioning that "[I] had the honor to deliver [my] last lecture at the Royal Institution," when he spoke about the possibility of a magnetic receiver being able to "work a recording instrument" (1905, p. 138). Doing so greatly increases the value of his system. Fortunately, "[Marconi] is glad to be able to announce that he has recently been able to construct a magnetic receiver that will work a relay and a recorder" (1905, p. 138). Besides having a more commercially viable system, Marconi can also reinforce that his past hypothesis from a previous presentation was correct, thus bolstering any new claims or hypotheses he makes for the future of his wireless telegraphy system. Marconi's ethos, therefore, is a necessary bridge for showing credibility in order to make convincing future predictions about the wireless's viability.

4 Expectations and Current Successes of the Wireless

The fact that the wireless was an advancement was not enough—Marconi had to show its practical necessity. Was the wireless simply a gadget that amateurs and inventors would tinker with in basements and sheds? Or would the wireless have commercial value that could be realized by society? One mission of the Royal

Institution and Royal Society of Arts was to promote applications of science. Marconi thus showed the wireless's economic value by illustrating instances where the wireless seemed crucial to social practices, often portraying the wireless as a useful humanitarian device. Marconi's wireless did not transform society; instead, it fit into social practice. Because telegraph and telephone wires were already in place, the wireless did not change communication needs: The need to communicate across vast distances already existed. But Marconi argued for the value of the wireless in three ways:

1. He created dissatisfaction with current technology, which speaks to the technology's perceived need.
2. He discussed the current successes and places where the wireless existed even if they were on a small scale.
3. He dispelled myths about the technology.

Marconi had to show dissatisfaction with current technology in order to make his new system seem necessary. New technology must work within the current system and fulfill a niche or demand, which was done rhetorically by a "relevant social group." Bijker's (1995) argument that "relevant social groups" help rhetorically (and physically) construct technologies is important to creating dissatisfaction. Although Marconi, the wireless's mouthpiece, spoke to a small audience, his words helped condition relevant social groups such as other scientists, engineers, business people, and the public. These groups had to believe that the wireless was important and then demand it or its services (telegrams, communication, rescue dispatching, and entertainment). Similarly, as Bazerman (1999) argued, Edison's marketing had to "create a dissatisfaction with a current technology" in order to induce consumers to purchase it (p. 142). Edison had to create dissatisfaction when he was developing his incandescent light bulb: He showed that gas lighting was inferior to his new electric lighting. In remote parts of the country the new technology meant the difference between light and darkness, but, because Edison was mainly marketing to urban areas using gas or candle lighting, he had to manufacture bias against non-Edison systems. Many advertisements in newspapers showed that "gas technology had a number of drawbacks": Besides the fear of fires and poisoning, the ads pointed out that electric lighting was more flattering to female beauty [Bazerman, p. 150].

As Edison competed with gas, Marconi had to compete with the technology of cable by showing the wireless as better and more versatile. Cables can only be used to communicate with the two countries, between which they are laid,

> but if a wireless connection is established between two such countries the stations may be instantly *used in time of war*, or in any other emergency, to communicate with other stations, situated say, at Gibraltar, the West Indies, or some inland point in North America, and also, if necessary, with war ships carrying apparatus tuned to the waves such stations radiate. (Marconi 1905, pp. 143–144, emphasis added).

Marconi tried to convince the relevant social group of scientists and engineers that cable had a limited use; also, he subtly warned that the wireless would be necessary if war were to break out. Juxtaposing international wireless communication and war can create an aura of insecurity, and his boast that the wireless was currently used "from Cornwall to the Scilly Islands, on the *not infrequent* occasions of the breaking down of the cables" further constructs cable technology as an unpredictable, unacceptable technology (1905, p. 144, emphasis added). Although Marconi did not lie about the problem with cables, he emphasized their faults to create dissatisfaction.

In addition to showing cable technology as less secure and less reliable than the wireless, Marconi seemed confident "in [the wireless's] ability to furnish a more economical means for the transmission of telegrams from England to America and from England to the colonies than the present service carried on by cables" (1905, p. 144). Therefore, Marconi quantified a "real" value in adopting the future wireless system: It was cheaper than the alternative. Finally, Marconi did not just argue that the wireless was inexpensive; he also proclaimed "that some day it would be possible to send messages to the other side of the globe," and that feat would allow "the possibility of carrying out tests of very great *scientific* interest" (1905, p.145, emphasis added). The wireless was not just a commercial instrument better than cable; it was a scientific development; it was progress understood as evolving the human capacity to communicate.

Unless relevant social groups can show a new technology, such as the wireless, is superior to current technology, it may fail. Marconi specifically creates dissatisfaction with another contemporary shipping technology/practice when, in his 1900 presentation, he discusses a test between the two vessels the *Alexandra* and the *Juno*. In order to create this dissatisfaction, he points out that wireless technology is superior to sight for signaling other ships. Apparently, the *Juno* transmitted signals from the flagship *Alexandra* to other ships with Marconi's wireless, which he called "the system" in this description, and "[t]his enabled evolutions to be carried out even when the flagship was out of sight" (1900, p. 291). Marconi went on to argue wireless progress and superiority over the current system by claiming "[communication among the ships] would have been impossible by means of flags or semaphores" (1900, p. 291). Semaphores are visual signals using flags, lights, or a person's arms to display an alphabetical code. His wireless system was superior to existing technologies, implying they are outdated and inefficient.

Besides creating dissatisfaction with current technologies, Marconi also shows that the wireless continues to be improved. Because technological narratives are ones of progress, pointing to "the old way of doing things" creates dissatisfaction. But instead of creating dissatisfaction about another technology, Marconi implicitly created dissatisfaction with older versions of the wireless. He did this effectively when he described tests done on commercial vessels. Because these tests showed improvement, they were inherently "progressive." Marconi provided the audience with examples of progress in order to construct a working image of the wireless. Throughout the wireless's invention and even after the radio became a black box, inventors and observers (for instance, assistants and reporters) stressed the range of communication applications. Marconi often referred to increased distances (for transmissions) in his presentations when discussing his wireless's range. Obviously, each development that increases range and quality can be accurately measured and easily reported. For example, Marconi used induction coils in tests in order to have signals travel farther than previous wireless apparatus. In March 1899, Marconi sent signals from "the South Foreland light-house and Wimereux, near Boulogne, over a distance of 30 miles" and "[t]he maximum distance obtained at that time…was 42 miles" (1900, p. 291). The July 1899 "evolutions" between the *Alexandra* and *Juno* had "messages [that] were obtained at no less than 74 nautical miles (85 land miles)" (1900, p. 292). This new induction coil or "tuned circuits" was the famous wireless patent number 7,777 (Hancock 1950/1974, p. 19). Marconi mentioned some specific tests of the new system and improved distances in his 1905 presentation to reinforce his technology's progress: While discussing yet another improvement he referred to a time in early 1900 when his system would only transmit 200 miles or less (p. 139); by 1902, Marconi showed that the system had greatly improved and messages could be "up to a distance of 2,300 miles" (1905, p. 140). Even if the audience knew about other wireless inventors (for instance, Bose, Deforest, or Tesla), Marconi's new patents were improving wireless technology. In other words, he created dissatisfaction with other wireless systems—even his own older devices—by mentioning he could now do more with his newer inventions.

Marconi expressed the benefits of the new improvements to his wireless in terms of modernist tropes like speed, efficiency, and even automation. This new receiver "is far more simple" and "requires far less attention.…But the chief advantage lies in the fact that with this receiver it is possible to attain a very high speed of working" (1905, p. 138). Marconi's improvement automated the recep-tion of transmissions, "[s]o for as speed is concerned…this new detector is not dependent upon the ability of the operator" (1905, p. 139). Therefore, he manu-factured dissatisfaction with older wireless devices. By automating the system (to some extent) and increasing the speed at which the system transmits signals, Marconi illustrated a more efficient system over the previous coherer device:[3] "In conjunction with Professor Fleming, [I have] recently introduced improvements which greatly increase the efficiency of the apparatus" (1905, p. 139). Marconi's focus on automation shows humans to be less efficient, thus, automation creates dissatisfaction with human involvement.[4] Efficiency meant the system was streamlined, reliable, and not wasteful; all of those qualities construct the wireless as a viable technology that fit with the time period's Fordist/Taylorist values, especially the perceived economic needs like shipping and, consequently, colo-nialism. The wireless brought the British Empire *closer*.

Marconi's overall goal for creating dissatisfaction with current technologies such as cables and visual signaling and older wireless technology was to foster an ideal representation of the wireless, one that speaks to technology's values. Obviously, efficiency and reliability are important for the technology to "catch on" or become part of the time period's reality. But Marconi's rhetoric manufactured the belief in a potential technology as a real artifact. The wireless, as a potential device, had to be made to fit the culture's value system. Simply being a "better alternative" to current technology was not enough to show that the wireless was valuable. Additionally, Marconi had to argue for its future potential. The next section demonstrates that the wireless's potential helped create its immediate value.

5 The Wireless's Technical Progress and Future Potential

Marconi often represented the wireless as a technology that had moved forward and continued to advance. Although Marconi briefly mentioned in one presentation that "the early development of [wireless telegraphy's] practical application was slow," the contemporary "practical application…is many times as great as the predictions of five years ago" (1900, p. 288). The new improvements he discussed in this early presentation allowed him to repeat that wireless telegraphy was beyond expectation: He could now "convey the intelligible alphabetical signals over distances far greater than had been believed to be possible a few years ago" (1900, p. 288). Therefore, the rhetoric of the wireless defined the invention, in part, as a technology beyond contemporary knowledge, a product exceeding expert expectations. Marconi also enhanced his scientific ethos during a description of his wireless's progress: He broke an assumed *law* (a black box) that related energy and wire height to the distance a signal may travel. Apparently, Marconi transmitted signals 85 miles, but "[a]ccording to a rigorous application of the law, 72 miles ought to have been obtained…thus we obtain a greater distance than the application of the law would lead us to believe" (1900, p. 289).[5] As he would eventually dispel the myth of wireless signals not being able to travel over the earth's horizon, refuting the mathematical law of wire height to wireless signal distance positioned the wireless as a technology not confined to conventional (or even scientific) wisdom. It exceeded expectations, the expectations of the scientific establishment.

Showing the wireless as "beyond expectations" gave it the special quality of constantly being a scientific/technological breakthrough. Each new wireless installment/innovation would seem to surpass the previous one. Such a quality helped Marconi's "damage control" when he recounts a problem with the Cape Breton station which had many accidents in 1903. Because of the accidents, "the [Marconi Wireless Signal Company, Inc.] decided not to attempt the transmission of any more public messages until such time as a reliable and continuous service could be maintained and guaranteed under all ordinary conditions" (Marconi 1905, p. 141). Although this could seem like a failure on face, success was just around the corner because "[a] new station, supplied with more powerful and perfect apparatus" was being built, and Marconi assured the audience "that in a very short time the practicality and reliability of transatlantic wireless telegraphy will be fully demonstrated" (1905, p. 142). He argued further that "[p]ending the reconstruction of these long-distance stations, valuable tests have been carried out, and daily commercial work is carried on over distances of about 2,000 miles" (1905, p. 142). Because the wireless had such a long reach, "five trans-Atlantic steamships are thus publishing a daily newspaper containing telegraphic messages of the latest news"

(Marconi 1905, p. 142). The wireless allowed passengers onboard ships to receive news while traveling just as railroad passengers received news via telegraph wires at different railroad stations (Bazerman 1999, p. 24). This idea of instant news fit the cultural desire to be in ready contact with others over vast distances.

Because the wireless was shown to be beyond expectations, Marconi could claim that his system would overcome current problems and improve. One issue he had to counter was the problem with "the daylight effect," a phenomenon that "may cut off [wireless] signals at great distances" just after day break (1905, p. 143). This effect is simply the fact that wireless signals do not need as much power to transmit in the night as they do in the day. Marconi allayed concerns by claiming he "does not for a moment think that this daylight effect will prove to be a serious drawback to the practical application of long distance wireless telegraphy" (1905, p. 142). Marconi did not give the audience a technical or scientific reason as to why he was not worried about the daylight effect on commercial operations, but his assurance might be good enough because he was the expert with personal technical capital. After all, he had proven the wireless to be a scientific breakthrough; believing the wireless will overcome "the daylight effect," which may not even be understood by the audience, is completely within reason.[6]

Besides reiterating that the wireless did more than it was expected to do, Marconi had to convince the audience that he actually sent wireless signals using transmitters and receivers outside of view from each other—beyond the earth's horizon. Marconi accomplished this, but, since it was contrary to scientific belief, he had to change the community's belief; he had to establish a new fact and dispel two myths: the myth of the supposed interference of the curvature of the earth and the myth of iron deposits affecting wireless transmissions. Many scientists did not believe signals could be sent beyond the horizon until after his messages crossed the Atlantic. He dispelled the idea that "iron in the hills" interfered with wireless communication by claiming such a belief is a "very unscientific opinion": He denied iron was a hindrance when conducting tests with "[a] fleet of 30 ironclads" or "transmit[ting] my messages with absolute success across the very high buildings of New York, the upper stories of which are iron" (1900, p. 295). This refutation added to Marconi's proclamation that the wireless put the "great force of nature at our disposal" (1900, p. 287) because nature—the curvature of the earth and iron—did not interfere with wireless signals. Marconi also dispelled myths by suggesting constant improvements mean that *the technology will work itself out*.[7] Similarly, in this context, the "daylight effect" appears to be a short-term hindrance that will work itself out because he "does not for a moment think that this daylight effect will prove to be a serious drawback to the practical application of long distance wireless telegraphy" (1905, p. 142). And the future proved his hypothesis to be correct.

The importance of dispelling myths about the wireless cannot be overlooked. As the wireless's mouthpiece, Marconi had to defend it from others professing doubt: "One of the chief objections which is raised against wireless telegraphy is that it is possible only to work two or a very limited number of stations in the immediate vicinity" (1905, p. 132). Marconi countered this objection by explaining his improvement in technical terms and argued that "[t]his objection appears to be much more serious to that section of the public which knows little or nothing of telegraphy in general than to telegraph engineers" (1905, p. 132); in other words, Marconi set boundaries for who could claim to have advanced wireless knowledge by saying, essentially, *we in the scientific community know better*, causing those who believe the myth to think twice. After all, who wants to be seen as unknowledgeable?

Marconi singles out another inventor to dispel myths as well. When describing his new syntonic system in 1901, he referred to experiments carried out by a certain Professor Slaby—a lesser known wireless inventor. Marconi dissected Slaby's supposed experiments by mentioning how he improved upon Slaby's work. Marconi even pointed to problems with Slaby's system by mentioning "[t]he reasons which demonstrate that a closed circuit, such as is employed by Slaby, must be a poor radiator, are obvious to those who have studied and read the classical works published since the time of Hertz's experiments" (1901a, p. 786). Basically, Marconi portrayed Slaby as ignorant with respect to wireless telegraphy fundamentals. Marconi, who was then the most prominent name in the field of wireless telegraphy, attempted to convince the audience that only non-engineers would believe the wireless could not be used with multiple stations in close proximity to each other.

But what is seemingly more important than the technical explanation of "syntonic apparatus" is Marconi's affirmation that "[i]t was possible nearly five years ago to send different messages simultaneously without interference" (1905, p. 136). Claims about that interference were, thus, historically inaccurate and at least five years out of date. Marconi showed a technical diagram of the "recent improvement introduced in method of tuning the receiver" (1905, p. 136), but, again, he referred to an unknowledgeable audience holding a mistaken belief. He claimed "[t]here exists at present among *the large section of the public* consid-erable misconception as to the feasibility of tuning or syntonizing wireless tele-graphic installations" (1905, p. 136, emphasis added). The issue he was about to address was that of intercepting messages. Although the reasons for Marconi addressing intercepted messages are not completely clear, we can infer that he may have been responding to a criticism of the wireless. "Syntonizing" refers to simultaneously sending and receiving wireless signals. Apparently, a small portion of the audience was concerned that wireless signals might not reach their intended destination if they happened to be picked up by unintended stations. Marconi corrected the audience—although his words were directed at "correcting" the public—about the "accepted understanding" of what constituted a truly 'inter-cepted' message: An intercepted message was not one that was picked up by the wrong receiver; it is a message blocked from the "intended recipient" (1905, p. 136). Marconi assured the audience that any wireless signals "tapped or over-heard at stations for which they are not intended…does not by any means prevent the messages from reaching their proper destination" (1905, p. 136).

This argument may seem strange to twenty-first century audiences. His assur-ance to the audience is not about security or privacy violations (or, for that matter, about identity theft). He claimed messages would be received regardless of whether another station "caught" them. Marconi's "assurance" would not be as effective with an audience more worried about eavesdropping than missing a message. Bringing up the issue of intercepted messages allowed Marconi the chance to counter the concern that signals could be maliciously blocked from the intended station. Although Marconi mentioned the possibility that "a powerful transmitter giving off strong waves...may prevent the reception of messages...the so-called interfering station is at the same time unable to read the message" (1905, p. 136), his main concern was to show "that any telegraph or telephone wire can be tapped, or the conversation going through it overheard, or its operation interfered with" (1905, p. 137). He added that "Sir William Preece has published results which go to show that it is possible to pick up...the conversation...passing through a telephone or telegraph wire" (1905, p. 137). There was no scientific reason for mentioning cables or telephone wires in this description, but, commercially, if the cable was susceptible to the same "tapping," Marconi's system was not inherently worse.

Additionally, Marconi also addressed the concern that multiple stations could interfere with wireless communications. He foreshadowed the future of the wireless by stating that

> [t]he ether about the English Channel has become, in consequence of great wireless activity, exceedingly lively, and a non-tuned receiver keeps picking up messages or parts of messages from various sources which very often render unreadable the message one is trying to receive. (1901a, p. 786)

By discussing the crowded "ether about the English Channel," Marconi planted the notion that the wireless was being used extensively. By itself such a statement could mean little, but, for a progressive technology such as the wireless, improvements were bound to develop. Marconi was creating dissatisfaction with his older system, which he had improved. He let the audience know that he was "now prepared with syntonic apparatus suitable for commercial purposes" (1901a, p. 786). Besides just being a *useful* device, the wireless must be shown to be a profitable one.

6 Applications and Economics

Marconi also rhetorically constructed the wireless as an economically viable technology. Any technology is necessarily influenced by economic factors: Technologies cost money to develop, make money through sales, and lose money (or not make money) by not fulfilling a niche. Marconi promoted the wireless as a technology useful for social practices and profitable for (current) investors. Marconi spoke about the wireless's profitability and potential for profit in forums with a history of using science applications to improve industry. Marconi's presentations spoke to these industrial concerns and created an image of a current and future technology with practical applications. In purely economic terms, Marconi manufactured both supply and demand for the wireless: He created it, and he fit it into current industries.

The wireless's early potential was evident in the description of its maritime applications and insurance savings. An important component to Marconi's early success was the interdependency of the Atlantic shipping fleet. Marconi's repeated sending of the Morse code signal for "S" across the Atlantic Ocean in 1901 worked, but only as a scientific event and not a commercial enterprise. It would be some time still before a reliable commercial transatlantic wireless service could be consistently maintained. No ship apparatus could transmit signals across the entire Atlantic Ocean in 1901 without a relay system. A wireless operator could transmit signals to any ship within range in order to have a message relayed. For instance, if a ship leaving Europe wanted to get a message to an American destination port (or any land station) shortly after leaving its port, the ship would have to contact another ship closer to the destination port; likewise, if a station in the United States wanted to contact a ship just leaving Europe, it would have to send signals to a closer ship that could then relay the message (Hancock 1950/1974, p. 46). Hancock noted that "[p]rogress brought improved apparatus and the need for such charts lessened" as new technology allowed ships farther wireless range (p. 47). The charts to which Hancock referred were schedules for ships entering and leaving ports on both sides of the Atlantic.

The relay system showed how Marconi's wireless fit into the current cargo shipping industry. Relaying was not new for communication and had been in use since the first telegraph wires. Early telegraph wires needed to have repeaters—stations set at certain distances that would repeat the message in order to send it farther along the wires—because the mechanisms to transmit signals were not powerful enough. Marconi's system was, therefore, based, in part, on current communication practices for weak output. Although this relay system eventually became obsolete, it established the Marconi system as a "standard" brand of wireless technology. By September 1901 "[t]he Marconi system was now in regular and continuous use on" forty-four ships from major shipping companies such as Cunard, Lloyd, and the Red Star Line (Hancock 1950/1974, p. 44). These vessels, combined with the 54 land-based stations on both sides of the Atlantic Ocean (Hancock, p. 44), made the Marconi system a well-known tool in shipping. By 1906 about 11 "ships were equipped with Marconi long-distance receiving apparatus capable of receiving messages throughout the whole course of their voyage across the Atlantic" (Hancock, p. 46). By 1907, when "139 ships, British and foreign, [were] fitted with Marconi's wireless telegraphy apparatus" (Hancock, p. 47), Marconi's system was already black boxed. But, before those commercial milestones, Marconi further constructed the wireless's importance for maritime usage by calling attention to its military application.

Marconi concluded one of his presentations by highlighting *his system's* benefits and future, specifically for the Royal Navy:

> As I have already stated, communication over a distance of 300 km is now being maintained with my system....It may be said that long distances of transmission are not necessarily an advantage, but I notice that the *navy* wants long-distance apparatus supplied to it. (1901a, p. 786, emphasis added)

Whoever may have said that transmitting over 100 km is not an advantage is unimportant; in fact, no one may have said it. By stating his system produced and received long-distance communications and mentioning that the military wanted this capability, Marconi created an ethos of popularity, which could increase interest in his company for military applications. Marconi went on to argue that his improved "25-centermetre-spark induction coil....might have been of use to the besieged garrisons in South Africa and China" (1901a, p. 784). However, Marconi did not promote combat operations as a good fit for the wireless overall, but, at times, he implied that the wireless had potential for combat. Early wireless tests by the US Navy (around 1910) were mainly communication tests that enhanced the scientific understanding of radio waves (c.f., Yeang 2004). Although war applications surround the wireless's rhetoric, it appears Marconi emphasized the wireless as an important safety feature for ships.

Marconi promoted the wireless by arguing that the world needed the safety it afforded: "[I]n the future...humanity is likely to have before very long to recognize in telegraphy through space without wires the most potent safeguard that has yet been devised to reduce the peril of the world's sea-going population" (1901a, p. 786). Goods were shipped between industrial nations, but raw materials were also shipped from colonies. High seas communication among different vessels for news and other correspondence made the wireless appear as a new modern convenience. Marconi concluded his 1900 presentation with the following prediction:

> [T]he progress made this year will greatly surpass what has been accomplished during the last twelve months; and, speaking what I believe to be sober sense, I say that by means of the wireless telegraph, telegrams will be as common and as much in daily use on the sea as at present on land. (p. 296).

That last speech not only stated the future potential of the wireless in commercial terms, but it also reinforced the notion that the technology seemed destined to "surpass" even the safety progress of the previous year. Therefore, because of the *progressive* nature of the wireless, the coming years would produce a technology for safety beyond expectations. This safety device seemed to be "right around the corner." Much would be reported in the 1910s about the safety of having the wireless at sea (for instance, rescuing those from the *Titanic*), but Marconi, at this point, rhetorically constructed the wireless as useful and life saving.

Marconi discussed commercial safety applications further by claiming the audience already understood the wireless's value and current use; for instance, Marconi highlighted that "the system has been in practical daily operation between the East Goodwin light-ship and the South Foreland light-house since December 24, 1898" by stating first "[a]s is probably known to most of you" (1900, p. 290). He went on to say that "[i]t is difficult to believe that *any person who knows* that wireless telegraphy has been in use between" the above locations "without breaking down on any single occasion, can believe or be justified in saying that wireless telegraphy is untrustworthy or uncertain in operation" (1900, pp. 290, 291, emphasis added). After all, as Marconi pointed out, the system was installed "in a small damp ship" and "under conditions which *try* the system to the utmost" (1900, p. 291, emphasis added), which seemed to suggest the question "Could you imagine how good the system would be under ideal conditions?"

In case the question does not pop into the minds of the audience, Marconi hoped that the government would invest more money into his system so "millions of pounds' worth of property and thousands of lives may be saved" (1900, p. 291). This call relates to the discussion about how important the wireless was for "avoid[ing] loss of life and property" (1900, p. 290). In one particular instance, Marconi showed how cost effective his system was by claiming, with "one short wireless message[,] property to the amount of £52,588 was saved" (1900, p. 290).

For the rest of its history, safety and savings would always be wireless attributes. Writing 50 years after Marconi crossed the English Channel; Hancock (1950/1974) claimed insurance underwriters as well as cargo transporters saved money because the wireless increased the safety and reliability of shipping (p. 16). Hancock offered an anonymous quote from an "authority" who claimed insurance companies may decrease "'premiums for insurance on Marconi-fitted ships'" (p. 31), which was what Marconi's allusion to "millions of pounds' worth of property" suggested. The discourse about safety and saving money was embedded in the wireless's rhetoric as an efficient and important system. Marconi spoke to the importance of the system because he was the wireless's mouthpiece, but he was also a proprietor who wanted a profitable system.

Marketing the wireless as a safe, efficient tool had nothing to do with the science of radio waves, and it did not have anything to do with the engineering behind its physical construction. Also, without marketing the benefits and superiority of the wireless over cable, the wireless would not have existed. Safety, dissatisfaction, and profitability of the wireless were independent of the science and engineering upon which the wireless was based. Profitability, though, is vital for the wireless's development. The Marconi Wireless Signal Company was incorporated to profit from the wireless technology it created. In a technology's life cycle there comes a point when "no one is necessary any more to shape the black box" (Latour 1987, p. 137). However, prior to becoming a black box, the wireless needed Maroni and engineers (or some relevant social group) to "shape" it into a reality. Latour believed that inventors usually handed over the maintenance of a technology to others to allow the technology "to more easily spread" (1987, p. 137). Although Marconi "handed over" some control when he sold his patent rights to other companies, establishing himself as the rightful owner of the patents was vital to continuing the wireless's profitability. In order to profit from the wireless, Marconi had to secure patent rights, and his discussion on those patents rhetorically constructed himself as the future "father of the radio."

7 Patents Showing Marconi's Ownership

When Marconi mentioned his patents by name in his presentations, he was claiming ownership for the wireless's improvements. Often when discussing patents, he used diagrams that experts would understand. These schematic representations—written in the language of electrical engineers—represent the various parts of the wireless system. Marconi referred to the diagrams in order to demonstrate how the various functions of the wireless operate. During his presentations he mainly explained what had changed with the new apparatus and did not go into great detail about the function, which is something a lay audience might appreciate. Even though the diagrams are the ultimate jargon of technical discourse—using symbols for coils, coherers, and other apparatus—Marconi's presentations are still accessible to non-electrical/electronic engineers, so researchers outside of the electrical engineering profession can uncover social influences in his work.

The 1901 presentation for the Society of Arts, London demonstrates the ways Marconi used rhetoric to assert his *ownership*. For instance, Marconi described how he "connect[s] the receiving aerial directly to the earth instead of to the coherer" by referring to a schematic description of "the new methods of connection which I adopted in 1898" (Marconi 1901a, p. 754) in order to do more than introduce the fact that he was about to describe the connection; he was also emphasizing *his* use and ownership of the improvement. Because he "adopted" this method three years earlier, he explicitly states that he was not discussing a theory but a fact. Immediately after he referred to the schematic, he established his history in the electrical engineering community by referring to "this improvement in the discourse delivered before the Royal Institution on February 2, 1900" (1901a, p. 755), showing that he has addressed the community about his improvement prior to the current presentation. He attempted to secure the idea of his legal rights to the new connection method by specifically mentioning "my first British patent specification referring [to the connection method] was applied for on June 1, 1898, No. 12,326, and published in due course" (1901a, p. 755). Marconi referred to this patent three more times in the presentation by always showing possession (i.e., "my patent," "my British patent," and "the description given by myself"). He also established that he had been working in the field considerably longer than most think: "It may probably surprise some of you when I mention how comparatively long ago some of the patents which I shall discuss to-night were applied for and perfected" (1901a, p. 754). This rhetorical signal tells the audience to be surprised that Marconi might not have been given enough credit for being in the field longer. Also, his technical ethos can only be bolstered by showing that he started in the field "comparatively long ago." These declarations may have little to do with the science behind transmitting wireless signals, but they definitely show ownership and build his ethos as an engineer in wireless telegraphy. Rhetorically, Marconi secured his property through patents and his status through invoking those patents.

In order to further his patent claims as *real* devices currently in use, Marconi offered the audience a practical success story of his system:

> If the system [of earthed wires] had not been used by me, I very much doubt whether we should have succeeded in maintaining communication with the East Goodwin lightship during 1899, in maintaining communication across the English Channel that same year…and in supplying the Admiralty in the course of the year 1900 with 32 installations. (1901a, p. 755)

Reminding the audience of his commercial successes showed ownership because he was not just showing the patent diagram; he was showing its practical use(s). Also, by using *ownership* language in regard to *his* success transmitting wireless signals, he, perhaps, boosted confidence in his potential commercial success. He legally, commercially, and rhetorically became *the* inventor of the wireless to the audience: Legally, he owned the patents; commercially, shipping fleets used his system; and rhetorically, the company, products, and stations had "Marconi" in their names.

Marconi cited his patent No. 12,326 to claim ownership of the "improvements on my original system, which have been in use by myself and my assistants for several years" (1901a, p. 755). Marconi states he is not new to wireless technology, but he also defended himself specifically against two others in the presentation: Professor Slaby, an inventor of a rival system, claimed in an article that in Maconi's system "[t]he receiving wire was suspended insulated and attached at the lower end to the coherer, the other pole of which was connected to earth," and G. Kapp claimed only in the "Slaby-d'Arco system...*the receiving wire is earthed*" (as cited in Marconi 1901a, p. 755). Marconi defended his claim that he used "earthed" receiving wires, a fact "openly discussed by the scientific press of this and other countries a long time previous to the date of Slaby's paper" (1901a, p. 755). He continued to prove he used earthed wires by citing other publications and a few diagrams, but his representation was not solely to show the makeup of the wireless; instead, he appeared to position himself commercially as the true owner of the wireless system. Even though he legally owned the patent No. 12,326, he went into considerable detail to prove the patent was part of his working system, a technology he had been using for quite some time.

Understandably, having a working technology as opposed to a non-working (or unverified) theory would lead to more credibility. Inventors had to worry about losing credibility for making outlandish claims. Edison himself overstated claims concerning his "discovery" of wireless transmissions in 1875, and "[h]e labeled the phenomenon the 'etheric force'" (Bazerman 1999, p. 27). Bazerman noted that Edison went to the press too quickly, causing the scientific community to claim his findings were "misguided and overblown," damaging his credibility (pp. 27–28). Marconi appeared to want to control how audiences perceived his system. He discussed his ownership of the system, but he also called attention to his contribution to science. In order to secure the scientific community's approval or acknowledgement of his ownership, Marconi impressed upon the audience the fact that others incorrectly understood his system. Of course, this idea related to a previous example mentioned above where Marconi claimed that only the unknowledgeable did not understand the "proper" workings of the wireless. Marconi's presentations were a mixture of scientific and technical knowledge: scientific in that he commented on the science of radio waves (passing around the horizon, unhindered by iron, signaling distance related to the height of a tower); and technical in that he described the apparatus (coherers, wires, balloons, transistors, etc.) that created those waves. The wireless was not yet a large-scale commodity in the early twentieth century. Marconi's audience may have been potential users, but most would not have been current users. His rhetoric prepared audiences to believe in the wireless and Marconi's stake in its invention.

Also, many readers/listeners would have been familiar with the work of other inventors. Marconi referred to Dr. Ambrose Fleming's lecture in his 1901 presentation "delivered before this society in November and December last year" (1901a, p. 755). His immediate goal was to explain that "electrical oscillations set up by the ordinary spark discharge method cease" because of "electrical radiation removing the energy in the form of electric waves" (1901a, p. 755). Apparently, Fleming, a prominent scientist at the time, observed "that in the case of conductors of a certain form the electric oscillations die away with great rapidity" (as cited in Marconi 1901a, p. 755). Marconi further built a scientific ethos by equating himself with Fleming. He, again, showed that he was "standing on the shoulders of greatness" when he claimed that he tried to "carr[y] out a great number of experiments by adding to the radiating and receiving wires inductance coils" similar to a patent held by Oliver Lodge (1901b, p. 781). Although Marconi was unsuccessful, he continued to forge ahead and was eventually able to improve his system and patent this improvement "by myself on March 21, 1900, No. 5,387" (1901b, pp. 781–782). Therefore, his improvement nullified the benefit of Lodge's patent.

Just as Lodge's patent was incomplete and untested, so was W. G. Brown's patent on the "use of two conductors of equal length joined to each side of the spark-gap" (Marconi 1901b, p. 782). Marconi mentioned Brown's work only to undermine its importance: "[Brown] did not describe the inductance in series between them and the spark-gap, which, according to my experience, is absolutely essential for long distance work" (1901b, p. 782). Marconi implied that long-distance communication was favorable, the way of the future. To show he was on the vanguard of wireless telegraphy, he also pointed out how he used certain techniques before others did. For instance, Marconi mentioned that "[t]he idea of using a Tesla coil to produce the oscillations is not new" in order to set up the fact that he had been using the Tesla coil in his commercial system since 1898 (1901b, p. 783). Although he mentioned that others such as Lodge and Braun used such a technique, he mentioned that "[his] idea was to associate with this compound radiator a receiver tuned to the frequency of the oscillations set up in the vertical wire by the condenser circuit" (1901b, p. 783). Thus, he separated his work from the others and further explained the technical details for using transformers and conductors to create these coveted oscillations.

Marconi shrewdly noted "that Professor Braun has recognized the necessity of tuning the circuits of the transmitter and receiver when using a Tesla coil in order to obtain syntonic effects, but I am not aware that such a proposal was published prior to the description given in [Marconi's famous patent No. 7,777]" (1901b, p. 783). The only reason to mention the fact that his system was the first to tune circuits when transmitting and receiving was to show his property, his patent. Although we can figure out the public experiment from his patent numbers, he described his patents as lab events, showing ownership and demonstrating his technical knowledge. The explanations are mostly devoid of real-world examples;

for instance, instead of saying where he accomplished or discovered that "it will not be difficult to transmit to any one of [the several stations], without danger of the message being picked up by the other stations for which it is not intended," he introduced this idea as a hypothetical possibility: "It is easy to understand that if we have several different receiving stations" (Marconi 1901b, p. 784). Although he tested the induction in a lab setting, Marconi's system in 1901 was not strained by simultaneous transmissions. In the future, simultaneous transmissions would be an issue, but they were not at the time of this presentation. Presenting the wireless in this way rhetorically constructed it as a technology that performs a certain way, but, physically, it was not quite a viable commercial system.

Marconi brought up another patent in his "concluding" section of the 1901 presentation to return to more criticism of Slaby's work. He cited Slaby's use of a "multiplicator" by claiming Slaby referred to "an especially wound induction coil ('induction-spule'), the function of which is to increase the electro-motive force of the oscillations at the end of the coherer" (as cited in Marconi 1901b, p. 785). Marconi told his audience that "[he] assume[d] that the multiplicator was an oscillation transformer performing the function of those described in my patent, dated June 1, 1898" (1901b, p. 785). Mentioning his patent here is an attempt to reinforce his property and accomplishment, but it also set up a dismissal of Slaby's experiment; apparently, Slaby claimed to have not used "a transformer, as [this multiplicator] has no secondary winding" (as cited in Marconi 1901b, p. 785). Marconi continued to dissect Slaby's supposed experiments by mentioning how he improved upon Slaby's work; at one point he even discredited Slaby's system by saying "[t]he reasons which demonstrate that a closed circuit, such as is employed by Slaby, must be a poor radiator, are obvious to those who have studied and read the classical works published since the time of Hertz's experiments" (1901b, p. 786). Basically, Marconi was saying that Slaby was out of touch with the fundamentals of the wireless, a claim he later repeated in his 1905 presentation that a certain wireless operation perceived to be a problem "appears to be much more serious to that section of the public which knows little or nothing of telegraphy in general than to telegraph engineers" (1905, p. 132). Marconi's rhetoric in the two previous quotations attempt to identify those who know and those who do not. He downplays problems by implying, essentially, that, if on believes a certain operation is a problem, that person is not properly educated in wireless telegraphy.

Marconi finished discrediting Slaby by telling the audience that he had a better working system that had overcome the "big" difficulties (at least as far as May 1901 was concerned; he was still seven months away from crossing the Atlantic):

Slaby has not yet described how to obtain different messages from transmitters situated at equal distances from receivers, which is much more difficult *in my experience*, nor does it appear possible with the method he describes to transmit various messages at the same time from one sending wire, as can be done with *the system I have just explained*. (Marconi 1901b, p. 786, emphasis added).

The legal ownership his patents secured, which is a different type of ownership from the social ownership of being associated with the wireless, granted him economic privileges and rights for implementing his system. Additionally, his presentations rhetorically *conveyed* his ownership and knowledge about wireless technology to a technical audience. Although a patent is a form of discourse that represents the wireless's physical nature, Marconi's presentations reified the idea of the wireless, bringing the physical (via patents and other diagrammatic representations), scientific, and economic aspects of the technology together. Also, the idea was *his* idea of the wireless.

Marconi's presentations show even in the most technical forum, he still spoke to the attitudes and values of the audience and emphasized the practical potential of his system and not simply the physical make up of the wireless. The presentations are examples of how technology is infused in the workings of a society. The government, with its seemingly endless coffers of tax revenue, needs to support large technological systems in order for these technologies to become realized. Marconi needed the British and Canadian governments to support (and allow) the construction of wireless stations and, eventually, hire him to install the wireless onboard naval vessels. Also, he needed important figures (such as the Queen, kings, dukes, the President, and others) to use the wireless to garner the public's attention and the support of policy makers. Again, technology is not created in a vacuum. Marconi needed governments to grant him not only the right to perform experiments on their soil, but also to finance part of his system and, eventually, a national infrastructure. In fact, even the Italian government "placed a 7,000 ton cruiser, the *Carlo Alberto*, at [Marconi's] disposal" to help him carry out some experiments (1905, p. 139). This assistance shows that governments are also *relevant social groups*, to borrow Bijker's phrase, in creating technology.

Although Marconi occasionally and softly criticized the British government for being behind the times, he immediately claimed "[I] considered it [my] duty to send the first messages [from Canada] to their Majesties the Kings of England and Italy, both of whom had previously given [me] much encouragement and assistance in [my] work" (1905, p. 141). Besides publicly thanking these two important figures, Marconi brought up the possibility of connecting different countries immediately through wireless communication by showing that "[m]essages were sent to His Majesty from Lord Minto, the Governor-General of Canada" and that "a message from President [Theodore] Roosevelt was successfully transmitted from this station [on Cape Cod] to His Majesty the King" (1905, p. 140). Having national leaders use new technology did more than legitimize the product; it caused the population, or at least a large portion of it, to recognize the technology as a force in their lives. Many presidents and state leaders have "opened" new technologies through ceremonies. Nye (1994) claimed that celebrating technology through ceremonies where politicians and other cultural elites gather creates a feeling within the population that the new technology, especially technology that shows off landmarks, "demonstrates how technological spectacle can produce bonds of solidarity" (p. 172). That ceremony manufactures a positive attitude for the new technology, which has been constructed in accordance with the culture's values.

Marconi's descriptions about implementing his system for satisfied customers were rhetorical techniques for creating a wireless consciousness. By discussing patents and commercial ventures, Marconi made the wireless a reality for the audience; knowing that the system was in operation somewhere or that "the admiralty are taking steps to introduce the system into general use in the navy" ("Wireless Telegraphy" 292) reified at least the belief in the technology because someone wais using it. The idea of the wireless, constructed rhetorically, existed before it became commercially viable: Marconi may have been building stations and equipping ships with wireless systems in 1900, but they were mainly for test purposes. After the wireless became a viable technology, it contributed to mass communication. Marconi's scientific and technical descriptions showed the benefits not just of a device, but of an international system of communication. Similarly, Edison was not just interested in inventing a light bulb; he wanted an entire electrical system that would guarantee households would be able to utilize the technology (Bazerman 1999, p. 159). Having a system as opposed to a simple tool makes the technology inevitable but only after the technology fits with current technology or practices. Unlike Aramis, the ill-fated mass transportation system that did not fit with Parisian commuter practices (Latour 1996), Marconi's system was easily installed on ships that were already crossing the Atlantic.

Marconi represented the wireless as an efficient, instant communication device and built the meaning of the wireless by juxtaposing it with the time period's cultural values. Efficiency and speed—two modernist tropes—speak to the values of Western industrialized nations, and they are values echoed in Marconi's presentations. Having royalty, celebrities, and presidents embrace these technologies makes the technologies seems like important cultural products that fit into the goals of the society. Fortunately, for Marconi and other inventors, progress is a value of industrial societies, so positioning a technology as an improvement over "the old way" fits the culture's ideology. The wireless had been rhetorically constructed, but it was not yet a realized technology; instead, it was being negotiated by relevant social groups. Marconi created an abstract concept of the wireless through rhetoric, therefore, creating the idea of a working, viable technology. Although I do not argue whether the audiences at the presentations accepted the wireless as real, I assume that Marconi's rhetoric was well received because his invention is recorded favorably by history. Which presentations convince whom is impossible to say, but the discourse surrounding the wireless is more than a marketing ploy: Marconi's language attempted to acclimate audiences to the idea of the wireless. The discourse, covering cultural pride, economics, ownership, and technical features, was evidence of the rhetoric of technical communication, which conveys ideas and meanings of technologies to various audiences. In the case of Marconi's presentations, he was speaking to technical and semi-technical audiences who gathered for similar events where distinguished scientists and inventors convened. Marconi's invitation to present and his Nobel Prize identify him as valuable to the scientific community.

To continue examining favorably representations of the wireless in the early twentieth century, the next chapter examines how journalists enmeshed in modernist values and practices represented Marconi's wireless in the popular press. Marconi's presentations, the popular press articles, and other types of discourse surrounding the wireless are technical communication situations.

References

Aristotle, (1991). *On rhetoric: A theory of civic discourse*. New York: Oxford University Press. (G. A. Kennedy, Trans.).

Bazerman, C. (1998). The production of technology and the production of human meaning. *Journal of Business and Technical Communication, 12*(3), 381–387.

Bazerman, C. (1999). *The languages of Edison's light*. Cambridge: MIT Press.

Bijker, W. E. (1995). *Of bicycles, bakelites, and bulbs: Toward a theory of socio technical change*. Cambridge: MIT Press.

Bucci, O. M., Pelosi, G., & Selleri, S. (2003). The work of Marconi in microwave communications. *IEEE Antennas and Propagation Magazine, 45*(5), 46–53.

Corazza, G. C. (1998). Guglielmo Marconi—Marconi's history. *Proceedings of the IEEE, 86*(7), 1307–1311.

Crowther, J. G. (1954). *Six great inventors: Watt, Stephenson, Edison, Marconi, Wright, Brothers, Whittle*. London: Hamilton Press.

Garratt, G. R. M. (2006). *The early history of radio: From Faraday to Marconi (IEE history of technology, no 20)*. Herts, United Kingdom: The Institution of Electrical Engineers. (Original work published 1994).

Gilbert, G. N., & Mulkay, M. (1984). *Opening pandora's box: A sociological analysis of scientists' discourse*. Cambridge: Cambridge University Press.

Hancock, H. E. (1974). *Wireless at sea*. New York: Arno. (Original work published 1950).

Hong, S. (2001). *Wireless: From Marconi's black box to the audion*. Cambridge: MIT Press.

Jaffe, A. (2009). Inventing the radio cosmopolitan: Vernacular modernism at a standstill. In M. Coyle, D. R. Cohen, & J. Lewty (Eds.), *Broadcasting modernism* (pp. 11–30). Tallahassee: University of Florida Press.

Jensen, P. (2000). *From the wireless to the web: The evolution of telecommunications, 1901–2001*. Sydney: University of New South Wales Press.

Kraeuter, D. W. (1990). The U.S patents of Armstrong, Conrad, De Forest, Du Mont, Farnsworth, Fessenden, Fleming, Kent, Marconi, and Zworykin. *AWA Review, 5*, 143–191.

Latour, B. (1987). *Science in action*. Cambridge: Harvard University Press.

Latour, B. (1996). *Aramis, or the love of technology*. Cambridge: Harvard University Press. (C. Porter, Trans.).

Marconi, D. (1982). *My father, Marconi* (2nd edn.). Ottawa: Balmuir.

Marconi, M. C. (2001). Marconi, my beloved (2nd edn.). In E. Marconi (Ed.), Boston, MA: Dante University of America Press

Marconi, G. (1900, Feb 2). Wireless telegraphy. *Smithsonian Annual Report, 1901*, 287–296. (Original published in *Proceedings of the Royal Institution of Great Britain, 16*(2), 247–256)

Marconi, G. (1901a). Syntonic wireless telegraphy. *Electrical Review, 38*(24), 754–756.

Marconi, G. (1901b). Syntonic wireless telegraphy II. *Electrical Review, 38*(25), 781–786.

Marconi, G. (1905). Recent advances in wireless telegraphy. *Smithsonian Annual Report, 1906*, 131–145.

McGrath, P. T. (1902/1999). A very loud electromagnetic voice. In R. Rhodes (Ed.), *Visions of technology: A century of vital debate about machines, systems and the human world*. New York: Touchstone.

Noble, D. F. (1999). *The religion of technology: The divinity of man and the spirit of invention*. New York: Penguin. (Original work published in 1997).

Nye, D. E. (1994). *American technological sublime*. Cambridge: MIT Press.

Tarrant, D. R. (2001). *Marconi's miracle: The wireless bridging of the Atlantic*. St. John's, Newfoundland: Flanker Press.

Weightman, G. (2003). *Signor Marconi's magic box*. Cambridge: Da Capo Press.

Yeang, C.-P. (2004). Scientific fact or engineering specification? The U.S. navy's experiments on wireless telegraphy circa 1910. *Technology and culture, 45*(1), 1–29.

Chapter 4
Popular Press Representations of Marconi's Wireless

> The development of some varieties of municipal engineering
> is absolutely dependent upon the development of public
> opinion and must proceed with it. The matter of street
> cleaning is largely a question of an improved public taste in
> the matter of street paving. Unless streets are well paved they
> cannot be well cleaned except at a prohibitive cost. To jump
> from one degree of cleanliness in this respect, to another,
> without a supporting public opinion, may be enough to wreck
> an administration and to set the tide of civic improvement
> running in the opposite direction. The newspaper is a great
> educator in these matters today.
>
> (Cooke 1915/1999, p. 61)

In the early twentieth century, Marconi's wireless excited journalists on both sides of the Atlantic. Although his invention was not yet commercially viable or universally accepted at the turn of the last century, many relevant social groups affixed positive meanings to it. After Marconi demonstrated the wireless's potential in his technical presentations and physical demonstrations, journalists reconstructed the wireless for larger audiences. Many set out to promote Marconi's "triumphs" in their periodicals. Even advertisers used "the wireless" to promote products: In the March 1903 issue of *World's Work*, an advertisement for Pears' soap boasted of "A Wireless Message Across the Atlantic… Sent 20 Years Ago" (Fig. 4.1). Positive representations of the wireless allowed readers "a glimpse into the future," just as Nye (1994) argued World's Fairs did because they displayed technology of tomorrow (p. 205). Descriptions of Marconi's invention and accompanying presentations appeared in periodicals as news events with much of the technical information removed or described differently for a lay audience; however, the popular press articles share several qualities of Marconi's presentations.

This chapter demonstrates how pro-Marconi popular press articles used tropes of progress to reconstruct the wireless in 13 articles from the following American periodicals dated from 1899 to 1905: *McClure's*, *World's Work*, *Current Literature*, *Frank Leslie's Popular Monthly*, *Living Age*, *North American Review*, and *The New York Times*. Although Marconi was not an American, these American popular press articles show how Marconi's experiments received international attention, which thrilled journalists and made them speculate about both the potential of the wireless and Marconi. This response is important because, though a British company in competition with other American and European inventors,

A. A. Toscano, *Marconi's Wireless and the Rhetoric of a New Technology*,
SpringerBriefs in Sociology, DOI: 10.1007/978-94-007-3977-2_4,
© The Author(s) 2012

Fig. 4.1 Pears' Soap advertisement (*World's Work* 1903, p. 3047)

Marconi and the Marconi Signal and Wireless Company emerged as *the* individual and company most responsible for bringing wireless technology to the world. History records Marconi as the "father of the radio," and American periodicals were one place where this status materialized. Their articles discussed similar international events and helped trace the process by which the wireless and

Marconi's status became embedded in the culture's collective conscious: These representations suggest his wireless fit with prevailing cultural values, attitudes, and practices of the early twentieth century. The articles mainly focus on Marconi, the man, his wireless feats and records, and his invention's future potential. While the wireless was not Marconi's invention alone, the popular press represented it as *his* invention, helping solidify his future title of "father of the radio." These reconstructions also showed the wireless as a progressive technology more efficient than contemporary alternatives. The wireless, consequently, reflected the social attitudes and values of the Industrial Revolution—specifically, the early twentieth century's *technophilia*. According to many journalists, the wireless would usher in a new communication standard and free humanity from physical wires or cables.

The following sections of this chapter discuss the importance of the popular press as a relevant social group, the instances where Marconi's persona is positively ascribed, the ways the popular press described the wireless as monumental, and the visions/prophecies of wireless benefits. First, the popular press is an important relevant social group because its texts represent the wireless's image and, therefore, its reality prior to becoming a black box. Second, the popular press constructed Marconi's trustworthiness through rhetoric, which helped build his scientific ethos and construct him as an important cultural figure. Finally, just as Marconi envisioned future practical applications for the wireless, journalists also reconstructed the wireless's value by discussing its potential. Interestingly, the popular press proclaimed that the wireless's potential was "beyond expectation" much more boldly than even Marconi.

1 The Relevance of Journalists and Their Descriptions of the Wireless

Before discussing the "relevance" of journalists, it is important to explain what is meant by "popular press," which distributed journalists' contributions. Journalists fulfill an important role as a relevant social group because they construct the news. According to Bazerman (1999), "[n]ewspapers are the daily world gathered into words and pictures" (p. 2). Bazerman discussed the importance of the "emerging world of news" for Edison's inventions: As technology allowed greater contact via telegraph communication, "journalism turned from small circulation partisan commentary to mass-circulation retelling of the happenings of the world" (p. 23). Journalists captured "[e]vents that happened anywhere… everywhere… or nowhere (like the promise of a new invention)" (Bazerman 1999, p. 23). I refer to the journalists' texts on Marconi's wireless as the "popular press" because I want to stress the *popular* nature of this particular venue. As Bazerman observed, the press "became a new kind of stage" for representing "technological advance[s]," one that popularized inventions and inventors (p. 23). Although I do not argue specifically how audiences received the wireless's popular representations, from journalist's descriptions of Marconi's wireless, I have found descriptions that adhere to prevailing cultural values. These accounts do not represent all wireless-based journalism; instead, they reflect how pro-Marconi descriptions conveyed the wireless's image as a new, efficient marker of progress.

Of course, Marconi's work was not confined just to newspapers: Other periodicals, such as magazines, also spoke to the mass public. The audience for popular press news was not an expert audience; instead, as Satia (2010) argued, the "lay press" allowed Marconi "to secure an alternative [to the technical press] source of legitimacy as a businessman and scientist" (p. 834). Satia focused on the rhetoric of warfare and empire in her analysis of popular press articles, which help "trace the close entanglement of radio technology with security in the public imagination, and Marconi's deliberate cultivation of that association" (p. 834). Because Marconi was successful in having the public associate him with the wireless, a rhetorical analysis does not need to attempt to examine exhaustively all popular press documents to draw valid conclusions about the discourse surrounding the wireless. There were (and, because of ongoing debates about Marconi's significance as the inventor over, for instance, Nikola Tesla, it is accurate to claim "are") other discourses that affixed meaning to the wireless. Although these could lead to alternative conclusions, they are not themselves *the* story; instead, readers should consider them alternative ways of perceiving Marconi's position as *an* important contributor to the rhetoric of the wireless. Because technology is not the sole creation of a single individual but the construction of a culture, Marconi's physical construction of his wireless is less important to a rhetorical analysis than surveying the positive discourse that rhetorically constructed the wireless as an important, monumental technology that fits Western ideals for progress.

While no discourse holds a technology's total value, texts from different genres reflect some of the cultural work a technology does. For instance, contemporary media report work on the supercollider, virtual reality systems, spacecraft for civilian travel, and even "wireless" Internet access. Also, commonly reported, not-yet-available scientific products are abundant in popular press reports on potential medicines/treatments. One need only recall that often repeated phrase about a medical breakthrough reported on the nightly news that is "pending FDA approval" to understand the parallel between the wireless prior to becoming a black box and a forthcoming medication or vaccine.[1] Pro-Marconi journalists did not need a consistently working wireless model to hype its potential to their audiences. However, they needed to be told the wireless was coming or in use somewhere in order to accept the technology. Fortunately, the public "loved" the wireless because the press promoted it, at least in part, in accordance with their values and expectations. No evidence exists that the public wholeheartedly believed Marconi's truthfulness or that the wireless was a physical reality at the time; however, the popular press discourse acted as a cultural repository where the wireless *existed* rhetorically.

All texts about technologies are sites for negotiation. Even though I do not argue specifically how or to what extent audiences *believed* in the popular representations, the articles I examine fit the early twentieth century's progressive industrialized ideology. Satia (2010), looking at how the popular press discourse created the idea for "mass communication," claimed "the press" did not identify the feelings of all the public, but "[the press] does disclose the scientific opinion and public information [the public] drew on to form their ideas during this period" (pp. 833–834). The articles I examine establish the journalists contemporary to Marconi as an important "relevant social group," and they had international reach. They described an exciting new scientific breakthrough—a "human" advancement. As Nye (1994) argued, technological achievements are important cultural markers that transcend nationality and excite or electrify the cultures in which they are invented (p. xiii). Although Marconi was an Italian subject doing most of his work for his British company, American journalists capture or report the excitement the wireless had on them to the public. The wireless bridged the Atlantic and made the world "smaller": It was not contained within a single society but spread from Britain to Italy to the United States to Hawaii to Japan and even to Africa. The international reach marked progress for the industrialized Western World, causing American periodicals to celebrate a foreign invention.[2] Even the contemporary technological/cultural critic Adams (1900/1974) "would have hugged Marconi" out of reverence for his invention as he stood awe-struck in front of the great dynamos at the 1899 World's Fair (p. 380). Marconi's status was not confined to Britain.

The emerging interest in science and technology in the early twentieth century is apparent from popular press descriptions, suggesting that scientific and technical subjects were of interest to wider audiences. In the previous chapter, Marconi's presentations were meant for technical or semi-technical audiences—people judging the physical viability of the wireless. Although non-scientists and non-technologists attended the Royal Institution and the Society of Arts lectures, they were not peer reviewers. The popular press coverage suggests what a culture (in general) found worthy to follow, and the wireless was important *news*. The articles were not for science and engineering discourse communities, but the journalists validated the wireless by reifying it in a popular forum. To popular audiences in the early twentieth century, *machines* were wonderful (even liberating) inventions. Later, however, many high modernist artists would lament the dehumanizing mechanization of consciousness and human labor. The popular press articles I examine contain no allusions to mechanical dehumanization, environmental degradation, or human alienation. Quite the contrary. Journalists provided glowing accounts and possibilities for the wireless; the only negativity is towards "outdated" technologies about which they created "dissatisfaction."

Although anti-Marconi articles exist, their attempt to dissuade the public from holding Marconi in high regard ultimately failed. In fact, one quasi-critical article about the wireless praised Marconi's "wonderful achievement of transmitting a message across the Atlantic," but claimed that tests "in the Mediterranean proved a 'total failure'" ("Wireless Telegraphy" 1902, p. 3). Nevertheless, the article does not conclude negatively. Instead, it connects optimistically about the new wireless technology because "[m]odern ingenuity is not daunted by the failure of experiments, but takes new courage from defeat, and triumphs in the end"

("Wireless Telegraphy" 1902, p. 3). Many popular press articles connect Marconi to this eventual "triumph."[3] What is important from the previous quasi-critical article is that it gives the wireless (as well as other modern technology) the status of being progress soon to come to fruition—it is destined to be part of the culture.

Journalists constructed Marconi as the main inventor in a group of important world-renowned figures: "Lodge, Fleming, Muirhead, Fessenden, de Forest, Tesla, Ducretet, Rochefort, Guarini, Popoff,[4] Arco, Brann, Slaby" (Waterbury 1903, p. 656). Mentioning Marconi alongside these famous inventors and scientists created an image of the soon-to-be "father of the radio": Marconi was among the great established scientific figures of the time period. Although one journalist commented that "in many cases [wireless] patents show that there have been independent discoveries of exactly the same thing in different countries at prac-tically the same time" (Waterbury 1903, p. 656), Marconi's importance as the main inventor was apparent when another article noted that "a dozen or more people have sprung up to share or to attempt to obtain the glory of originality," but "Marconi stands unquestionably at the head of the list" ("Recent Wireless Telegraphy Development" 1903, p. 419). Although another story of the wireless unfolds through the many patent-battle articles, I focus on descriptions of Marconi's wireless's events and the excitement associated with it because the legal construction is not as valuable for this study of rhetoric aimed at a popular audience. While rivals had their own national and international attention, these other systems did not receive the same *buzz* contemporarily or historically. The popular press was an important relevant social group that constructed Marconi's system as a benchmark even though he "compiled" his apparatus from his own work and the inventions of others. Other systems and apparatus existed, but they existed as just that—*other*—or other-than-the-benchmark standard according to the popular press.[5] The articles described Marconi's invention's profitability, usefulness, scientific importance, and improvement over cable—the existing "inefficient" technology of the time period. They build the rhetoric of the wireless.

2 Marconi's Apotheosis by the Popular Press

The popular press constructed Marconi as an almost royalty-like celebrity. In 1897, the King of Italy bestowed upon Marconi the title *chevalier*, "corresponding to an English knighthood" (Friend 1902, p. 529). Much later in his career he received the Italian nobleman title of *Marchese* in 1929, and "King George V conferred upon him the Honorary Knighthood of the Grand Cross of the Victorian Order" (Hancock 1950/1974, p. 163), which established his cultural significance. His pseudo-royal status began almost immediately after his first public experi-ments. Marconi's apparatus, even though not commercially viable in the early twentieth century, had famous users, adding to its status as an exciting, important invention. The popular press reported on wireless transmissions between the Queen of England and her son (the future King Edward VII), and later the press reported messages between Theodore Roosevelt and King Edward VII. One article reported that transmissions between the Queen and Prince "were the memorable tests... between Osborne House, on the Isle of Wight, and the royal yacht, with the Prince of Wales aboard" (Moffett 1899, p. 4). Such descriptions invoke the wireless as an invention for royalty. Satia (2010) argued that "[b]y courting royalty, Marconi burnished his image as an agent of empire and illustrated his technology's fitness for use by the imperial state" (p. 846).

The press's identification of Marconi as "royalty" helped lend importance to his wireless system. The wireless was a newsworthy event, which meant Marconi, the man who brought the wireless to the world, had celebrity-like attributes. Furthermore, these reports on Marconi's status reinforce his magical character. After all, sending invisible waves was magical for this early twentieth-century audience.[6] Other inventors also had magician-like qualities in the popular press. Leo Henricus Arthur Baekland, who invented "the first truly synthetic plastic" (Bijker 1995, p. 101), was described by the popular press as a "'grand duke, wizard, and bohemian'" (Bijker 1995, p. 197). Although Bijker argued that the "wizard" attribute distorted the social construction of a technology by possibly portraying the invention as "the development of... the genius of" an inventor (p. 197), the magical aura described recurs in popular press accounts, especially those accounts that identify a single inventor of a new technology. Edison was popularly known as the "wizard of Menlo Park," and Marconi similarly was often attributed with magician-like qualities. For instance, one article specifically describes Marconi's wireless's ability to transmit invisible waves through the "ether": "Electric waves cannot be seen, but electricians have learned how to incite them, to a certain extent how to control them, and have devised cunning instruments which register their presence" (Baker 1902a, p. 8). Therefore, Marconi, the electrician whose system was being described, can manipulate nature as a magician.

Marconi made the invisible a reality through his "cunning" apparatus and magical quality. One article, written by an author credited as "Friend" (1902) who writes a glowing account of Marconi's accomplishments as well as a brief history of the wireless up to March 1902, described Marconi as a "wonder-worker" and "an audacious experimenter," which, combined with the idea that he is "a man who has largely complemented whatever promises he has made" (p. 530), suggested his limitless potential as an inventor. Andrew Carnegie established that Marconi's persona and power were beyond the audience's imagination: "'No one can tell or even dream of what tremendous things he will be able to do in a few years'" (as cited in Wallace 1902, p. 1). While commenting on Marconi's transatlantic feat, one article established that even though the world "accepted that [Marconi] was now preparing for bolder ventures... none imagined a project so amazing as he entertained" (Friend 1902, p. 530).

These popular press accounts, therefore, established the wireless—or, more appropriately for 1899–1905, the science of radio waves—as a reality; Marconi created or discovered knowledge, and "to the world, what was hardly a probability three months ago is now an *undisputed fact*" (McClure 1902, p. 526, emphasis added). That previous comment came from an article reporting on a new advancement circa April 1902 and wass explicitly stating a "truth" for the audience: Marconi had discovered, created, and established a "fact." As with all technical communication, an audience must believe in the legitimacy of the technical information conveyed—the source must appear credible. In order for Marconi's wireless to be considered an important technology, the audience must trust Marconi. Popular press accounts of his trustworthiness and future potential both built his technical/scientific ethos and built the wireless, rhetorically, as a viable technology. Communicating an inventor's trustworthiness is underscored in an article that claimed, "at present one certainly takes the public into one's confidence when one sends a wireless telegraphic message" (Waterbury 1903, p. 658). Wireless signals are invisible—they cannot be seen or heard. In contrast, the public can see or hear "the construction of ships, the transmission of sound, the detonation of explosives, etc." (Waterbury 1903, p. 658). Because wireless signals are intangible, discourse must *construct* their existence, so Marconi's celebrity and pseudo-royalty status created his image, which, in turn, helped establish the wireless's potential for the audience.

The popular press reporting a transmission as "fact" further added to the excitement surrounding the wireless. Because industrial nations had a growing scientifically and technically aware population, a population consuming technical communication in a popular forum, journalists created almost fantastic stories about new technologies to excite readers. The fact that so many popular press accounts about Marconi's experiments exist is one testament to the excitement his wireless created. One journalist remarked that "the world throbbed with the surprise of his [crossing the Atlantic], and the cables were loaded with congratulatory messages to him" (Friend 1902, p. 532). Additionally, periodicals would preview up-coming experiments or interviews with inventors alongside advertisements for adventure novels and other sensational bestsellers to be printed: *McClure's Magazine* advertised between a review of Robert Louis Stevenson's *St. Ives* and Anthony Hope's sequel to "The Prisoner of Zenda" that a future issue would have an interview with Marconi and J. Chandra Bose.[7] The fascination and awareness of important new technologies was not new in the early twentieth century. World's Fairs had been exciting people for decades. But machines had a *superhuman* quality to them, and many popular press accounts insisted the audience ought to trust a technology over a human. In fact, journalists portrayed Marconi himself as a machine, a characteristic that allowed him to succeed: "[T]he keynote to [Marconi's] success is his unfailing industry and energy. He is a human dynamo" (Friend 1902, p. 532). Dynamos were the computers of the time period; therefore, calling Marconi a "dynamo" makes him superhuman, a more evolved individual.

But accolades would not have had the same force in print had Marconi not been seen as trustworthy. Years of reports and even contact with wireless apparatus might have reified Marconi's "miracle," but journalists helped construct the wireless by filtering Marconi's credibility. The popular press built Marconi's scientific ethos by praising his great scientific/technical mind and built trust by pointing to his careful nature when revealing success to the public. Edison also developed public credibility, allowing him to claim he had a working light bulb nearly a year prior to a completed prototype "and an additional year before a full system was ready" (Bazerman 1999, p. 13). Bazerman argued that "the meanings people attributed to [Edison] were embedded in specific and well-developed systems of communication that made his work seem credible" (p. 13). Therefore, Edison had to have an aura of credibility to help establish his invention as extant or, simply, possible. As was often the case for Marconi, his credibility established that the invention was "just around the corner." Belief is enough for a technology to exist rhetorically. Marconi's wireless's rhetoric necessarily established his invention, for the popular press argued his truthfulness and dependability, suggesting that he would continue to improve his invention.

Many popular press articles touted Marconi's record for not making bombastic claims about experiments and his potential for future development. One article claimed Marconi and the wireless share the following "confidence" for future progress: "[W]ith an assured achievement and practical daily working of 200 miles, and an experimental success of 2,000, it is *beyond* dispute that Marconi's work warrants the *confidence* which enthusiasts have in *its future*" (Friend 1902, p. 533 emphasis added). Another article claimed that Marconi's truthfulness "unquestionably carried great weight in convincing Mr. Edison, Mr. Graham Bell, and others of equal note of the literal truth of his" experiments (Baker 1902a, p. 7). Major S. Flood Page (Marconi's assistant) reaffirmed Marconi's credibility in the popular press by stating that Marconi "has never made any statement in public until he has been absolutely certain of the fact: he has never had to withdraw any statement that he has made as to his progress in the past" (as cited in Baker 1902a, p. 7). While reporting Marconi's claim that it would be possible to "establish wireless communication between San Francisco and Manila without an intermediary station at Honolulu or Wake Island," the journalist pointed out that "Mr. Marconi has never been known to say that he could do a thing which he either had not already done or very shortly did do" (Wallace 1902, p. 4).

Marconi's trustworthiness was also constructed as "business success." The popular press (re)constructed the wireless as destined to work out its flaws, with a final product beyond the readers' imaginations. Similarly, one article assured readers that Marconi would fulfill his promise of creating a wireless station in Nantucket that would allow for consistent, reliable transatlantic transmissions: "It is the first prophecy that Marconi has made since he began his work in wireless telegraphy seven years ago. He has not failed before. Few believe that he will fail now" (McClure 1902, p. 527). In effect, the article attempts to recruit the audience to believe in Marconi's invention. In April 1902, when that article was published, Marconi's work was mainly experimental, but it existed as a viable idea with the article's reassurance that Marconi spoke the truth and could not fail. This and other reports helped establish Marconi's trustworthiness and helped establish the wireless as a "real" technology, partially by using Marconi's personality and celebrity status to build the wireless's image as an actual technology progressing to better levels, evolving into a monumental achievement.

3 Popular Press Representations of the Wireless as Progression of Past Science

Much like Marconi built a scientific ethos by showing himself as continuing the important work of past scientists, the popular press often showed Marconi's work as a continuation, a progression, of past science. In one article, the author recognized a simple version of "wireless telegraphy" existed before Marconi, "for a score of inventors had preceded him… and Heinrich Hertz, the famous German savant, had proven that" wireless signals could be sent (Friend 1902, pp. 531–532). But the article went on to claim Marconi kept persevering "and spent five years before he solved the [early wireless] difficulties" (Friend 1902, p. 532). In other words, Marconi continued from the point where others had stopped. Such an account contributed to the "lone inventor myth" because it never named other inventors from whom Marconi compiled various components to use in his wireless apparatus;[8] it simply mentioned the scientists who speculated about wireless signals. This omission adds to the ahistoricity of technology (an important component of Marinetti's (1909/1971) "The Founding and Manifesto of Futurism," discussed in the next chapter): Although journalists discussed past science, they did not usually write about the many other specific contributors to the invention. Marconi was often given credit for the *entire* wireless in popular articles; journalists perceived him as being the most important inventor because he created a commercial enterprise for the wireless. Therefore, the popular press accounts were as much about Marconi, the man, as they were about the wireless.

Several authors of this time period—literary, popular, and artistic—portrayed machines as all powerful. While describing the power required to send wireless signals, one journalist showed his awe of the telegraphic spark by claiming it to be "as thick as a man's wrist… the most powerful electric flash yet devised" (McClure 1902, p. 526). The journalist also claimed "the very ground near by [the spark] quivered and cracked with the energy," and "[n]o human being could stand near the huge coil which produced this tremendous flash of lightning" (McClure 1902, p. 526). The article showed the "power" and "energy" needed to transmit signals across vast distances. Therefore, the image the popular press created is one that positioned the wireless as real and monumental. These larger-than-life accounts of the wireless praised Marconi and his invention, a powerful machine crackling with energy.

Machines also added validity to Marconi's experiments. After observing Marconi's transmissions from land stations to the *Philadelphia*, which were captured by a recording tape, the same journalist argued for the transmission's authenticity because "[w]hen a machine does a thing, we humans believe; so long as a man stands between, we doubt" (McClure 1902, p. 525). That is, humans distrust what they perceive as human inefficiency; a machine, in contrast, is trustworthy and accurate because "there [is] no human agency to 'think' or 'imagine,' and perhaps err" (McClure 1902, p. 525). Another writer also argued that machines were better "from the fact that a record on a tape is quite independent of what any man with a telephone thinks he hears" (Waterbury 1903, p. 660). Even Marconi's system, which "was successfully established among the Sandwich Islands... failed through unreliability caused [solely] by lack of skill among the native operators" (Waterbury 1903, p. 661). The machine, the wireless, was not to blame—the *human* operators were. While Henry Adams' account of the awesome power he felt from the dynamos suggests how many in the late-nineteenth century worshipped machines, these popular press accounts reveal that some believed technology to be superior to human interpretation. Machines were seen as more accurate and, therefore, more useful. Such a view reaffirms dominant progress tropes of modernism such as speed and efficiency associated with technology. The wireless was a new marker of human progress. The popular press used exciting language to describe the wireless and attempted to "prove" the wireless's importance by discussing its current usage and "obvious" future potential.

The biographical literature describes Marconi as a genius for being the "first" to create a viable technology. After envisioning the wireless's potential, which came to him after he read an article by his former professor Augosto Righi, Marconi set out to make a practical, commercially viable invention (Hancock 1950/1974; Tarrant 2001). However, though many popular press accounts described past scientists who developed theories about radio waves, journalists favorable to Marconi begun with him as the inventor of the first *practical* apparatus; and it was this practicality that privileged Marconi's status as *the* inventor of the wireless. In fact, even technical journalists overlooked Marconi's contemporaries: In an article from *The Electrician* (1898), the author solidified Marconi's preeminent status as the wireless inventor:

> For some considerable time the scientific aspect of this development (Hertz and Clerk Maxwell) completely obscured its more *practical applications*. Scientists were so charmed with the experimental evidence it afforded as to the validity of Maxwell's electromagnetic theory, that for many years the fact that these experiments possessed any practical value as a means of signalling between two pieces of physically, mechanically disconnected apparatus almost escaped their notice... *All honour is due to Marconi for having been the first* to bring prominently forward before official bodies and the public the possibility, and, indeed, *the eminent practicability of using Hertzian waves* for telegraphing between two places not connected by an electrical conductor. (as cited in Hancock 1950/1974, pp. 6–7, emphasis added)[9]

The science might have been the work of others, but the practical invention was Marconi's.

Although Marconi appeared as the sole inventor, articles also suggested Marconi's modesty. Many articles established that Marconi "makes no claim to being the first to experiment along the lines which led to wireless telegraphy... it remained for Marconi to perfect a [wireless] system and put it into *practical* working order" (Baker 1902a, p. 8, emphasis added). One journalist noted that many other inventors helped contribute to the wireless's invention but argued that "Marconi's discovery gave the clue to practical and useful ranges; he was the first to see the commercial value of Wireless Telegraphy, and the Marconi Company was the first company in the field to exploit the new discovery" (Waterbury 1903,

p. 656). Several pro-Marconi popular press articles agreed with the above ones but made it clear to the readers that "[Marconi] deserves all the credit the world can give him for making the idea of such great practical value" (Wallace 1902, p. 2). These journalists praised his business acumen, which was explored further in Marconi's biographies, and embedded "usefulness" into the rhetoric of the wireless.

Occasionally, eminent scientists who would write about Marconi's status in popular periodicals would not ignore other contributors as journalists often did, but they still touted Marconi's importance. Professor John Ambrose Fleming, Marconi's friend and mentor, minimized the contributions of Marconi's contemporaries and portrayed Marconi as the inventor who had the vision for commerce. Professor Fleming (1899), later Sir John Ambrose Fleming, traced the "world's" knowledge of ether as a conductor[10] from "Prof. J. C. Maxwell," who "left as his most splendid intellectual legacy to the world his remarkable Electro-magnetic Theory of Light" (p. 632). He noted that this electromagnetic science was "research of the present century" (p. 631), a point Marconi borrowed in his 1900 presentation (discussed in the previous chapter). Fleming covered Clerk-Maxwell, Faraday, and Hertz—the past electromagnetic scientists—in detail, but he simply lists Marconi's contemporaries: "Lodge, Fitzgerald, Trouton, Savasin, de la Reve, Bose and many other physicists extend[ing]" Hertz's research (p. 636). Fleming portrayed Marconi as the *major* wireless technologist "[s]tarting from [the scientists'] known facts" (p. 636). Marconi's experiments marked the beginning of what Fleming called "the Ether Age" (p. 640): This new age marked human progress, and Fleming, writing in May 1899, foresaw the twentieth century as a period of time history would judge according to the wireless's ability to produce practical applications associated with "the ether waves we are learning to employ" (p. 640).

Interestingly, Marconi's crossing of the Atlantic appears to be the end of historical discussions of past scientists in the popular press. This situation may be due to the fact that the wireless's Atlantic crossing solved an important question: Could wireless signals make it around the Earth's horizon? The only exception to this omission was in a June 1902 article in *McClure's Magazine* that briefly discussed Clerk-Maxwell and the fact that Hertz stumbled upon his own theory while "trying to disprove Clerk-Maxwell's theory" (Wallace 1902, p. 1). The article even listed contemporary inventors who tinkered with Hertz's experiments but argued "it remained for the young Italian (Marconi) to jump across the gap of years of scientific study and make *practical* the most important discovery since Faraday invented the induction coil" (Wallace 1902, p. 2, emphasis added). Other inventors such as Varley, Calzecchi-Onesti, and Branly created imperfect coherers based on Faraday's coil, but Marconi perfected those gadgets and "began then a series of experiments which would have discouraged any less determined man" (Wallace 1902, p. 2). Although Hertz, Faraday, and other non-contemporary "electromagnetic" scientists are mentioned in glowing accounts of Marconi, their inclusion still does not diminish Marconi's status. Other articles attempted to "shed light" on contemporary inventors, but rarely did they try to supplant Marconi as *the main* wireless inventor.[11]

Being seen as the sole inventor was part of Marconi's charisma, and charisma is important for inventors being perceived as trustworthy and capable. Bazerman (1999) argued that "Edison was the charismatic center of the organizations that formed around him" (p. 259). Bazerman went on to argue that Edison's "first institutions to develop, manufacture, and disseminate his system of delivering light and power were built on the force of his authority and the trust he granted his close associates"; his charisma, therefore, built the foundation of his corporate legacy (p. 259). Bazerman argued from Max Weber's theory of charisma as an aspect of authority: Followers place their "'personal trust in the leader's revelation, his heroism or his exemplary character'" (p. 260). The authority in the relationship of chief inventor to assistant inventors mirrored Marconi's authority in the popular press. The articles promoted Marconi as trustworthy, which acted as proof of the wireless's future applications and current existence. Marconi most likely had charismatic power over his close associates, but the popular press used his words—journalists often quoted Marconi in their articles—to construct the wireless's potentiality for their audiences. Marconi, the inventor, was as important in the popular press accounts as his invention.

The popular press helped rhetorically construct the wireless's potential black box status (the future radio) by presenting its monumental nature. Long before DC Comics' Superman excited audiences with his super powers, Marconi "leaped the ocean at one single bound" (Friend 1902, p. 529). His crossing of the Atlantic inspired writers to portray him as a "latter-day Columbus, with the same splendid daring, [who] faced the broad expanse of waters to demonstrate another *mighty scientific fact*" (Friend 1902, p. 529, emphasis added). Fleming (1899) commented that "[p]ublic attention, on both sides of the Atlantic, has recently been strongly directed to the possibilities of telegraphy through space, by *remarkable experiments* of Sig. G. Marconi" (p. 630, emphasis added). Fleming wrote just after Marconi crossed the English Channel, but he prophesized the potential of Marconi's work as "the crowning achievement in a long series of scientific labors" (p. 631). Another journalist believed the wireless had yet to reach its full potential, but even a limited invention would still be good: "Let his system be limited to but one hundred miles, and within that radius it will develop *inestimable services*" (Iles 1902, p. 1785, emphasis added). The image constructed is that of an inventor and invention unbounded by any constraints. The audience is supposed to assume that the technology would be beneficial beyond their expectations. The journalists wrote about Marconi's wireless as if it were a black box, but it was not even viable for transmitting across the Atlantic at the time on a regular basis; however, rhetorically, it signaled human progress.

Marconi mainly conducted scientific experiments during this early period; he was still "tinkering" with his apparatus and mainly trying to boost its transmission capacity. Even his crossing of the Atlantic was a kind of tinkering but on a scale much larger than when he conducted his first experiments "transmitting messages over distances of a few miles" back in Villa Grifone, the enormous estate near Bologna, Italy where Marconi grew up (Tarrant 2001, p. 22). With the Atlantic experiment, though, his international tinkering made news, and journalists rhetorically constructed the wireless through their accolades. One article praised the wireless (specifically, the crossing of the Atlantic) as unimaginable and "of a nature to balk human credulity" (Baker 1902a, p. 4).[12] The article's introduction claimed 12 December 1901 is "a day destined to be important in the annals of invention" (Baker 1902b, p. 299). The praises continued by noting Marconi's work was "[o]ne of the great wonders of science ever wrought" and that Marconi had delivered "a new scientific wonder" to the world (Baker 1902b, p. 299). Another article proclaimed "[t]here can scarcely be a reader of *McClure's Magazine* who has not known of the great work which Marconi has accomplished in wireless telegraphy" (Wallace 1902, p. 1).[13] The journalist told readers that skeptics should not doubt Marconi's triumph: If skeptics were present at the wireless demonstrations, "Marconi would be hailed by them, as he has been hailed by many great scientists who have seen his system in operation, as one of the greatest inventors of all time" (Wallace 1902, p. 1). *The New York Times* (1901) also hailed Marconi and his Atlantic crossing as monumental: One article claimed Marconi was a giant, and "his name will stand through the ages among the very first of the world's great inventors" ("The Epoch-Making Marconi" p. 8); and months later after a land-to sea experiment, *The New York Times* claimed the occasion "marks the beginning of an epoch in international intercommunication" ("Wireless Telegraphy" 1902, p. 6). Another article addressed the lay audience by noting that "[t]o those of us unfamiliar with electrical apparatus there is something *incomprehensible* in the feat of Marconi" (Iles 1902, p. 1784, emphasis added). These praises rhetorically constructed both Marconi and the wireless as groundbreaking.

4 Popular Press Constructions of Military and Commercial Uses of the Wireless

Much like Marconi's rhetorical representations to the scientific and technical community, the popular press focused on more than the technical descriptions of the wireless. On face, the lack of journalists' expertise in wireless science may appear to have been the obvious reason for non-technical descriptions. However, not only were these descriptions forms of technical communication, but many popular press accounts also included technical descriptions along with current successes, future potential, profitability, and other discourse common to Marconi's own presentations. One possible reason for this was that technical descriptions meant nothing to a popular audience without the knowledge of the wireless's commercial use (or potential). The wireless could not simply exist as a physical product; it had to *mean* something to the audience to whom it was described; it had to be constructed as useful. The popular press descriptions suggest the audience's imaginations would be excited by new breakthroughs and future potentials. Consumers could not yet buy radios, but they learned about wireless transmissions on the high seas and "Marconigrams"—wireless telegrams—being sent by important people.

The articles in this chapter discussed the "countless" uses of the wireless, projecting its importance and potential. For instance, one article listed the possible uses of the wireless when commenting on its future profitability:

> [T]here will be considerable profit from leases of the service to great lake steamers, to pleasure yachts, and to a number of ships in our merchant marine. In addition, there is to be considered the possibility of Government use; the operation of the system overland, along our coast, in the lighthouse stations; by the Weather Bureau in foretelling approach of storms; in time of war by our armies; and on the ships of our navy. (Wallace 1902, p. 4)

The popular press discussed wartime applications as much as other potential commercial applications. The growing militarization of the West, foreshadowing WWI, appeared on the pages of the popular press accounts, and the wireless became an important technology for future wars. Satia (2010) noticed that Marconi "present[ed] his technology as something that could bind an empire that was being pulled apart by the cable companies' punishing prices—a rending that was particularly disconcerting during a time of war [in South Africa]" (p. 833). The industrial West was quickly becoming a highly militaristic group of nations, and they pursued military technology through WWI and WWII, and that drive even continues through today (e.g., remotely piloted drones).

Marconi spent some time on the wireless's wartime potential in his presentations, but these popular press articles speculated on such applications (maneuvers and surveillance) much more. The growing agitation of European nations, building towards global wars, and their warlike past could have unconsciously influenced journalists covering new technologies. Eventually, F. T. Marinetti would figuratively describe the wireless as a weapon assisting Italy's campaign in Tripoli in 1911. The wireless's war potential described by journalists acts as a rhetorical appeal in these accounts. One such article warned that "[w]ithout secrecy no system of wireless telegraphy could ever reach great commercial importance" (Baker 1902a, p. 10). The article summarized Marconi's response to the wireless's secretive capabilities by discussing a "reflector" that "could be faced in any desired direction" and, therefore, would not allow an unintended ship to pick up the signal (Baker 1902a, p. 10). That description was not enough for the author who pointed out that "an enemy [ship] might still creep in between the sending a receiving stations" (Baker 1902a, p. 10). Such conjecture at an early stage of the wireless shows at the very least a popular suspicion of other nations or acceptance of war and the new technologies that may be used in future wars.

Concern with wartime use appeared in other popular press accounts. An interviewer asked Marconi about message secrecy and if during wartime "communications between battleships or armies [would] be at the mercy of any one including enemies" (Moffett 1899, p. 12). Marconi's position was that the wireless would allow the U. S. "to keep close guard over Havana harbor without sending" in the navy; instead, the U. S. would need only "a single fast cruiser" if "[the] Americans were at war with Spain" (Moffett 1899, p. 16). That article from June 1899 came less than six months after the Treaty of Paris ended the Spanish-American War, thus, showing the relevance of potential future wartime wireless applications. Many popular press accounts cover the potential for war, and, because the article was an interview, Marconi's words created the wireless as a potential device for "[t]he warfare of the future," which "will have startling things in it; perhaps the steering of torpedo craft from a distance will be" possible (Moffett 1899, p. 17). Such advancements would have been considered fantastic to a world without automobiles; in fact, bicycles were the most "hi-tech" mode of personal transportation in 1899 even though internal combustion engines were being developed.

Other popular press accounts constructed the wireless's versatility for wartime uses: The wireless "can be moved as easily as a machine gun... and keeps in communication bodies of troops within four days' march of one another" (Lyle 1905, p. 5845). To emphasize the war-like nature or, at least, potential of the time period, one writer even asked how "the nations—huge belligerent individuals that they are—" will use the wireless (Lyle 1905, p. 5847), suggesting the audience expected or was receptive to descriptions of a technology's wartime application. The journalists also constructed the wireless as an important tool for colonization. The United States had just "won" its first Pacific territory during the Spanish-American War—Guam. Because cable would have been outrageously expensive, one article claimed America's interests in the Pacific would be improved by "the installation of [Marconi's] system between California and Honolulu and Manila, thus joining together by *invisible* links, as it were, the widely scattered *possessions* of the United States in the Orient" (Wallace 1902, p. 1, emphasis added). The journalist suggested that these American "possessions" had to be in contact with the U. S. mainland in order for America to prosper in its new imperialistic endeavors in the Pacific. Demonstrating that the wireless supported imperialism implies that the wireless will assist in military expansion. Furthermore, the journalist told the audience that America's "enormous telegraph business with Europe; our trade with Cuba, Porto Rico, the Philippines, and Alaska makes us more and more of a world power," and Marconi's wireless would become "the daily means of communication from our shores to our various *possessions* and other countries over seas" (Wallace 1902, p. 3, emphasis added). In the British press, war potential and imperial cohesion were factors that led to supporting Marconi's technology (Satia 2010, p. 849).

The popular press not only described the wireless's current applications but also predicted the wide range of future possibilities. Just as Marconi described his system's current usage during his presentations, the popular press focused on the wireless's current commercial applications and often used exact figures for costs. Articles on the wireless dispelled the notion that "wireless telegraphy is still largely in the uncertain experimental stage" by noting "it has long since passed from the laboratory to a wide commercial use" (Baker 1902a, p. 11). The press emphasized the speed at which Marconi advanced: Wireless "development... has been astonishingly rapid," and "[m]ost of the ships of the great navies of Europe and *all important* ocean liners are now fitted with the 'wireless'" (Baker 1902a, p. 11, emphasis added). One article told the audience that they "may communicate with a friend on almost any transatlantic liner... send... money, or give notice of legal action"; also, according to the article, one could even play a game of chess with" people on different ships or in different countries (Lyle 1905, p. 5843). The wireless supported industrialization and kept the "businessman" efficient: "A passenger on an ocean grayhound no longer loses a week" because "[the passenger] learns that the Russians are retreating in good order," and "he notes the wheat crop reports from Argentina, and straightaway orders his New York broker to sell or buy" (Lyle 1905, p. 5843).

Other articles report the wireless's profitability by noting that "[e]ach installation on a transatlantic passenger ship now nets about $5,000 a year," and "more than 8,000 words were transmitted in a few hours from an ocean liner, and on another ship the receipts for two days' operation amounted to $300" (Wallace 1902, p. 3). The article went further into detail by claiming "this business is increasing in volume" and the future profit estimate "from one installation... would amount to over a million dollars a year" (Wallace 1902, p. 3). Another article specified that wireless telegrams (marconigrams) cost "twelve cents a word for transmission to [seagoing] vessels" and that the "[Marconi] company was now actually doing a profitable business on a commercial basis" (Baker 1902a, p. 11). In other words, profits make an invention real because that means "useful" and, therefore, "in demand." The legitimacy or accuracy of these profit estimates are irrelevant to the image they construct. Rhetorically, the wireless was profitable, and people appeared to be using it at an increasing rate, which meant it would be more profitable in the future because "progress" was embedded in the rhetoric of the wireless.

Showing Marconi as the origination of commercial applications established his credibility for the claims he and the popular press articles made about the wireless's future potential. Marconi made the "impossible... later come to pass" (Lyle 1905, p. 5848). Because Marconi made this wireless "miracle" possible, his claims about future possibilities were more believable than those of a non-celebrity inventor. One "fantastic" potential the popular press reported was the Internet-like qualities possible (ones not recognized for over 90 years): "the news might be ticked off tapes every hour right into the houses of all subscribers" with Marconi apparatus (Moffett 1899, p. 16); and "the time will even come when the great banking and business houses, or even families and friends, will each have its own wireless system" (Baker 1902a, p. 11).[14] Marconi's credibility helped construct an image of the wireless's future potential, and his business sense also seemed important for journalists: One journalist shows Marconi as a man in league with Edison and Bell, and "[b]ecause of these facts of character, and the thoroughly business-like manner in which the plans have gone ahead for the various utilities of his system, it is difficult to know just where to limit Marconi's possibilities" (Wallace 1902, p. 4). Such articles claimed Marconi's system had arrived, furthering the image of the wireless as a permanent technology. Also, the fact that Marconi's English and American companies had capital backing and his stock had been rapidly acquired "is sufficient evidence of the confidence in which the *investing public* holds Marconi and his work" (Wallace 1902, p. 4, emphasis added). This "investing public" was confident that Marconi would continue to be profitable; otherwise, as the popular press implied, he would not have excited the business-oriented groups. Once the rhetoric of the wireless attached profitability to its meaning, the technology became a good investment because it was profitable (or appears to be). Such circular reasoning might seem fallacious as a purely logical argument, but the rhetoric of its profitability, constructed by discourse, is no different than economic ideas about the psychology of investors. Confidence in the market leads investors to invest in a market they feel is profitable.

These faceless, nameless investors and the general audience might have been fascinated by the thought of what was to come next in wireless technology when the popular press claimed improvements seemed always to be "on the way." Articles often quoted Marconi or spoke through his persona, explaining that he would overcome current obstacles. Marconi made a bold statement shortly after crossing the Atlantic to one journalist: "Give me a week at Nantucket and I will guarantee to receive signals from England… and all kinds of messages across the Atlantic" once he has built a reliable station (as cited in McClure 1902, p. 527). He also claimed his experiments in syntony—tuning transmissions for specific frequencies—were not perfect because "the electric tuning of a particular transmitter to a particular receiver" still permitted some interference, but this "is a possibility in the future," which "bids fair soon to be realized" (as cited in Moffett 1899, p. 16). Another journalist claimed being able accurately to warn "vessels entering a dangerous harbor in thick weather… is one of the developments of the near future" (Baker 1902a, p. 10). The popular press, especially those favorable to Marconi, helped construct the idea of a working technology. And the Marconi system, specifically, would help usher in "the time when messages would be regularly flashing between Europe and America," a time "much nearer than most people realized" (Baker 1902a, p. 11).

An interesting contrast to the above rhetorical claims that the wireless was sure to accomplish its goals appears in an article from *The North American Review*. The article supported many of Marconi's claims and maintained that the wireless was an extremely important invention, but it also claimed that the "importance lies more in the future than in the present[,] for Wireless Telegraphy is still in that nebulous state which prevents one from fairly judging whether or not it is of such real value as its present condition would indicate" (Waterbury 1903, p. 655). The article differed from the other positive Marconi articles in its discussion of the merits of many other wireless systems; for instance, it briefly mentioned the de Forest system—the biggest American rival—as well as the various "national" systems that tried to emerge in France, Germany, Austria, Russia, and Spain (Waterbury 1903, pp. 661–662). The article merely reviewed which systems the different European countries happened to use. Unlike the other popular press articles I have analyzed, which are obviously pro-Marconi, this one did not argue for future potential of one system over another. It even promoted the polygenesis theory of the wireless's invention: "It is the old story over again that is found in the history of so many inventions: the world being ripe for the idea, the minds of many men in many countries were turned to it at the same instant" (Waterbury 1903, p. 656). Such a statement advanced the notion that the wireless was a socially constructed technology as opposed to the invention of a solitary individual. Therefore, the article attempted to minimize Marconi's role by being somewhat skeptical of the wireless's future and by not focusing as much (or solely as other accounts do) on Marconi's particular system. However, it pointed to Marconi as the first to see the commercial potential. It just did not emphasize Marconi's system as *the* system; instead, it argued that the future of the wireless looked promising. The more pro-Marconi articles argued specifically for Marconi's future potential and current success; also, those articles show that Marconi's system would replace the current inefficient cable technology.

The popular press created dissatisfaction with cables by pointing out their shortcomings—unreliability and outrageous expense. Cables were the established technology for long-distance communication: Telegraph wires had been in place for nearly 60 years and the transatlantic cable had been operating for nearly 40 years before Marconi sent wireless signals across the Atlantic. Just as Marconi created dissatisfaction with cable technology, the popular press constructed "wired" communication as unsatisfactory. One article noted that "Professor [Michael I.] Pupin"—a Serbian emigrant and physicist who created an important induction coil for telegraph wires—"has pointed out that… any one cable cannot work faster than a single Marconi installation," and "it is evident that the Marconi system has all the elements of competing successfully with the cables now in greater operation" (Wallace 1902, p. 3). Other popular press accounts go into greater detail on why cables were not as good; for instance, one article suggested "that [Marconi] would be able to build and equip stations on both sides of the Atlantic for less than $150,000″ (Baker 1902a, p. 11). In contrast, "[a] cable across the Atlantic costs between $3,000,000 and $4,000,000, and it is a constant source of expenditure for repairs" (Baker 1902a, p. 11). Furthermore, "messages which now go by cable at twenty-five cents a word might be sent profitably at a cent a word or less" (Baker 1902a, p. 12).[15] Consumers would "win" by being able to send cheap marconi-grams, and investors or other interested business parties (e.g., the government) could establish a Marconi wireless station for a fraction of the cost of laying cable.

Many popular press accounts prophesized the wireless's replacement of cable. Satia (2010) found that Marconi created doubt in the cable companies and, therefore, made his system seem superior by pointing to "punishing cable rates" and "promis[ing] to end colonial isolation" (p. 844). One article created dissat-isfaction with cable by claiming offshore lightships provided "the Marconi system admirable opportunity of replacing cables, which are very expensive and in con-stant danger of breaking" (Moffett 1899, pp. 13–14). Speaking in an interview about savings with the wireless, Marconi claimed that "deep-sea cable costs $750″ and landed cables cost about $1,000 per mile (as cited in Moffett 1899, p. 16). Marconi also mentioned "the great expense of keeping a cable steamer constantly in commission making repairs and laying new lengths" (as cited in Moffett 1899, p. 16). On the other hand, Marconi claimed "[a]ll we need is a couple of masts and a little wire," and "[t]he wear and tear is practically nothing" (as cited in Moffett 1899, p. 16). The expensive, unreliable cables could not withstand the inexpen-siveness and versatility of the wireless. The popular press created dissatisfaction with cable technology, which, in turn, constructed more than the wireless's potential; dissatisfaction made the wireless a *real* alternative. Rhetorically, the wireless was cable's replacement—the new, better technology.

And alternatives were important for the hyper-industrialized nations of the early twentieth century. The time period favored new technologies, and, judging by how much emphasis Marconi and the popular press placed on progress, the wireless had to fit that ideal before it would become realized. Of course, the *professional* relevant social groups, such as engineers, scientists, and businesspeople certainly played a beginning role in physically constructing the wireless and other technologies. However, the public were potential users who also had to be convinced of the wireless's worth. The image the popular press constructed was that of an important new technology. For instance, the wireless could be used during war "over stretches where it might be impossible for the telegraph corps to string wires or for couriers to pass on account of the presence of the enemy" (Baker 1902a, p. 12). The popular press even claimed "it is beyond dispute that Marconi's work warrants the confidence which enthusiasts have in its *future*" (Friend 1902, p. 533, emphasis added). Because of Marconi's "miracle," in the near future "[t]he Pacific cable will be unnecessary[,] [t]he Atlantic cables will be abandoned[, and] [l]and lines eventually may be forced out of business" (Friend 1902, p. 533). Because the popular press engaged in such dissatisfaction rhetoric, the wireless's image—at least in the popular press articles—became a real alternative. The popular press probably would not have written so many articles had they not thought the public would be engulfed or moved by them. Using tropes of progress, the articles boasted that Marconi's invention brought the "future" to the present. Often the "beyond belief" or "around the corner" descriptions constructed the rhetoric of the wireless. It was almost here, and it would be better than one thought it could be.

And the wireless, which was ultimately successful, adhered to the same strategies as the other two discourses I consider: Marconi's work stands for progress. Rhetoric constructed the wireless as a progressive, efficient, and profitable technology and constructed Marconi himself as a genius inventor who could be trusted to bring the future to the world. One could certainly examine the accounts of other wireless inventors' systems to understand how they, too, were rhetorically constructed. What is important in this chapter is that Marconi's work was being closely followed by the popular press and (re)presented according to particularly important themes: The wireless was a monumental new invention based on nearly 100 years of research, important for war, commercially viable, and more efficient, therefore, better than cable. Such themes as those above suggest the audience would be receptive to the wireless because it fit the "modern" lifestyle socially, economically, and even personally. The popular press created the simultaneous images of a technology that's "right around the corner" and also "here and now." Those seemingly contradictory images have a rhetorical basis: The popular press accounts portrayed both the current uses for the wireless and prophesized its future. Therefore, the wireless was "the technology of the day"; older technologies, then, became outdated. Creating dissatisfaction with cable and other "outdated" technologies seemed to be a key component of the values associated with wireless technology: In contrast to cables and wires, the wireless was cheaper, more versatile, more efficient, and more modern.

In the next chapter, I analyze how an artistic movement (re)presented the wireless along with other early twentieth-century technologies in order to give a more robust analysis of the wireless's rhetorical representations alongside other *progressive* technologies. I do not argue that the wireless caused artists to glorify technology; instead, technologies were already being fetishized or invoking awe (c.f., Adams 1974; Nye 1994). As an industrial product of modernity, the wireless speaks to the theories of modernism. F. T. Marinetti, the founder of the first avant-garde (Futurism), further built the rhetoric of the wireless through his aesthetics. Instead of being a scientist, engineer, journalist, or consumer, Marinetti was a lover of the wireless—a modernist *technophile*—who shared a passion for efficiency with the likes of Frederick W. Taylor and Henry Ford.

References

Adams, H. (1974). The dynamo and the virgin. In E. Samuels (Ed.), *The education of Henry Adams*. Boston: Houghton. (Original work published in 1900).

Friend, A. (1902). Marconi, the man. *Frank Leslie's Popular Monthly, 53*(5), 529–533.

Baker, R. S. (1902a). Marconi's achievement: Telegraphing across the ocean without wires. *McClure's Magazine, 18*(4), 4–12.

Baker, R. S. (1902b). Marconi's achievement. *Current Literature, 32*(3), 299.

Bazerman, C. (1999). *The languages of Edison's light*. Cambridge: MIT Press.

Bijker, W. E. (1995). *Of bicycles, bakelites, and bulbs: Toward a theory of sociotechnical change*. Cambridge: MIT Press.

Cooke, M. L (1915/1999) Paving propaganda. In: R. Rhodes (Ed.), *Visions of technology: A century of vital debate about machines, systems and the human world* (pp. 60–61). New York: Touchstone.

[Sir] Fleming, J. A. (1899). Scientific history and future uses of wireless telegraphy. *The North American Review, 168*(510), 630–640.

Hancock, H. E. (1974). *Wireless at sea*. New York: Arno. (Original work published 1950).

Iles, G. (1902). Marconi's triumph. *World's Work*, pp. 1784–1785.

Lyle, Jr. E. P., (1905) The advance of 'wireless.' *World's Work*, pp. 5842–5848.

Marinetti, F. T (1909/1971) The founding and manifesto of Futurism. In R. W. Flint (Ed.), *Marinetti: Selected writings* (pp. 39–44). New York: Farrar, Straus and Giroux (Flint R. W., Coppotelli A. A., Trans.).

McClure, H. H. (1902). Messages to mid-ocean: Marconi's own story of his latest triumph. *McClures's Magazine, 18*(6), 525–527.

Moffett, C. (1899). Marconi's wireless telegraph. *McClure's Magazine, 13*(2), 4–17.

Nye, D. E. (1994). *American technological sublime*. Cambridge: MIT Press.

Pears' Soap (1903). [Advertisement]. *World's Work*, p. 3047.

Recent wireless telegraphy development. (1903). *Current Literature, 34*(4), 419.

Satia, P. (2010). War, wireless, and empire: Marconi and the British warfare State, 1896–1903. *Technology and Culture, 51*(4), 829–853.

Tarrant, D. R. (2001). *Marconi's miracle: The wireless bridging of the Atlantic*. St. John's, Newfoundland: Flanker Press.

Wallace, H. (1902). A great American enterprise: Development of Marconi's inventions in the United States and it dependencies. *McClure's Magazine, 19*(2), 1–4.

Waterbury, J. I. (1903). The international preliminary conference to formulate regulations governing wireless telegraphy. *The North American Review, 177*(564), 655–670.

Wireless Telegraphy. (1902). *New York Times*, p. 6.

Chapter 5
Tropes of Progress in F. T. Marinetti's Early Futurist Texts

> *History is more or less bunk. It's tradition. We don't want tradition. We want to live in the present and the only history that is worth a tinker's dam is the history we make today.*

(Ford 1916/1999, p. 61)

Western industrialized societies of the early twentieth century were entrenched in ideologies favoring progress: From human advancements to management science, citizens in the industrialized West during the early twentieth century believed "efficiency" was one of the highest goals of progressive life. Taylor's (1911/1967) efficiency manifesto *The Principles of Scientific Management*, for example, articulated efficiency principles that were put into practical use by one of the leading industrialists of the time, Henry Ford. Fordism/Taylorism exemplifies the early twentieth century's fascination with efficiency and progress, and modernist artists captured similar images, attitudes, and situations related to these values in their works. After all, art reflects the artist's world and worldview. A world inundated with technological change became a muse for some and just referents to others. European and American avant-gardes contributed to new artistic movements that incorporated and promoted new technological advances. A machine gun, an automobile, and a wireless telegraphy station replaced (somewhat) "traditional" muses such as an ocean, a mountain range, and a flower. Even artists who did not glorify technological progress were affected by the new tools of science that excited the minds of artists and audiences.[1] To live in industrialized countries meant one could experience the *progress* around him or her. New advancements were integral elements of avant-gardes such as Futurism.

This chapter demonstrates how F. T. Marinetti incorporated values and attitudes of progress associated with the wireless and other early twentieth-century technologies into his early Futurist manifestos. His manifestos are products of a period consumed by new scientific, technological, and industrial discoveries and, therefore, reflected the culture's overall belief that "progress" was an important social goal. For Marinetti, the wireless and other contemporary technologies signified humanity's progress (ion) away from an irrelevant past.

A. A. Toscano, *Marconi's Wireless and the Rhetoric of a New Technology*,
SpringerBriefs in Sociology, DOI: 10.1007/978-94-007-3977-2_5,
© The Author(s) 2012

Other artists were not so sure new technological advancements deserved such praise; for instance, D. H. Lawrence, Virginia Woolf, H. G. Wells, and Aldus Huxley presented technology in less exalted and more skeptical ways; they did not subscribe to technological utopian dreams. Although D. H. Lawrence had an interest in Futurism, he considered the movement immature, and his works did not promote technology as any savior. Whether Lawrence (1920/1995) had coal mines cut into the English landscape or created characters dehumanized by "modern" technology and industrialization, *Women in Love* critiqued the modernist reconstruction of technology as progressive—in the sense that *progressive* connotes beneficence or necessity. Although different modernist artists did not share the same views on technology, technology still had a role in modernist texts.[2] In her novel *To the Lighthouse* Virginia Woolf (1927/1989) broke up her narrative with asides about events in the war that referenced destructive technologies: "A shell exploded. Twenty or thirty young men were blown up in France, among them Andrew Ramsey, whose death, mercifully, was instantaneous" (p. 133). Lawrence and Woolf were critical of technology because of its destructive, alienating qualities.

In contrast, Marinetti fetishized technology and promoted new advancements as beneficial products and markers of civilization. Marinetti described these technologies similarly to how Marconi and the popular press described the wireless but in exaggerated ways. Using tropes of progress in his early Futurist texts, Marinetti praised how technologies exemplify the values of speed, efficiency, evolution, and ahistoricity. Those values manifested themselves into rhetorical constructions of technologies through various types of discourse. His texts portrayed early twentieth-century inventions as saviors of humanity, the pride of industrialization, and a model for humans to emulate. According to Marinetti's Futurist agenda, technology would help humans "escape" from *passéisme*, which is a love for the sentimental past and not for the industrial future. Also, as an avant-garde artist, Marinetti's texts displayed an experimental aspect through *telegraphic lyricism* or mimicking telegraphic discourse.

My analysis is meant to be not an exhaustive study on Marinetti or Futurism but an analysis that suggests Marinetti's fascination with technology indicates that he is a product of modernity. Futurism is often used interchangeably with Italian Futurism, which mistakenly implies Marinetti's aesthetics were universally accepted by the various Futurist groups—Russian Futurism, Cubo-Futurism, Ego-Futurism—and that his aesthetics were representative of all avant-garde aesthetics. Although Marinetti was not the only avant-garde artist (just as Marconi was not the only inventor of the wireless), many scholars note, as the first avant-garde, Marinetti influenced all of the avant-gardes that followed (Bondenella and Bondenella 1979; Butler 1994; Kirby 1971; Perloff 1986; Poggioli 1968; Rainey 1998). Bondenella and Bondenella (1979) observed that "[Marinetti's] ideas were received enthusiastically by most of the principal writers of the times," but later many withdrew their support or denied Futurism's influence (p. 316). Also, the avant-garde was not the only crowd receptive to Marinetti.[3]

To demonstrate how Marinetti's aesthetics related to industrial or technological tropes of progress, I analyze the following texts from his early phase of Futurism:[4]

1. "The Founding and Manifesto of Futurism" (1909/1971),
2. "The Birth of a Futurist Aesthetic" (1911a/1971),
3. "Electrical War (A Futurist Vision-Hypothesis)" (1911b/1971),
4. *Destruction of Syntax-[Wireless Imagination]-Words in Freedom* (1913/1973),
5. and *Zang Tumb Tumb*, (1914/1987).

I cover the last text *Zang Tumb Tumb* briefly in order to give some examples of Marinetti's *parole in libertá* poetry.[5] Much of Marinetti's aesthetic license is obvious in the first four texts, but *Zang Tumb Tumb*'s onomatopoeic style and avant-garde nature are best *seen* and not described. Readers should examine Richard Pioli's (1987) translation of Futurist texts *Stung by Salt and War* and Elizabeth R. Napier and Barbara Studholme's (2002) translation of Marinetti's texts *Selected Poems and Related Prose* (compiled by Luce Marinetti, Marinetti's daughter) in order to observe the multiple typefaces, font sizes, and styles that make Marinetti's *parole in libertá* unchained. These foundational texts of Italian Futurism and, therefore, the historical avant-garde were published shortly after Marconi's wireless became a black box. Marinetti's reliance on the machine in his art is more than an allegory of speed, efficiency, evolution, and the *future*; he also pushes Italian modernization.

To demonstrate Marinetti's avant-garde penchant for glorifying technologies in his manifestos, I discuss how Marinetti's texts position him as a product of modernity: His nationalism, accompanied by a desire for militarism and hyper-industrialization, fit the cultural values and practices of early twentieth-century industrial nations. Furthermore, technologies were muses for Marinetti, and his texts showed science fiction fascination with machines. Whether he was describing imagined futuristic monoplanes or "twenty-first-century" wireless warfare, Marinetti constructed another type of favorable rhetoric in order to explain the (potential) usefulness of an invention. In his poetry and manifestos, he echoed the other discourses that described the wireless as "beyond expectations."

1 Marinetti's Response to the "Cult of Efficiency" Surrounding New Technologies

The wireless was one of these important technologies that performed "beyond expectations" in Marinetti's work. However, in contrast to the exaggerated claims about the wireless in the popular press, Marinetti's (re)construction of the wireless in more a science fiction conception. His polemics rarely focus on concrete possibilities for new machines; instead, he made bombastic claims about an unrealistic future, a utopia brought about through technology. He also represented technologies by describing them as exciting or fantastic tools that would be economic and industrial "saviors." He often groups the wireless together with telephones, automobiles, machine guns, and even airplanes when claiming "inventions" would bring Italy to greatness. Such nationalist sentiment emerged when he advocated war and attacked what he considered weakness in a past-loving culture.

Marinetti believed techno-salvation was possible and issued the first avant-garde manifesto in 1909 to show the explicit importance of technology. For Marinetti, technologies did not just advance progress; they were the muses which humans and his new Futurist movement needed to emulate. The word "Futurism" implies that Marinetti's goal was for the future: He wanted Italy to advance technologically away from the past, which was dead to him, and move into a new industrial age, one that favored the machine. By advancing mechanistic "virtues," Marinetti proposed that life—not just art—ought to follow technological values: Humans should be fast like cars, explosive like bombs, super productive like factories, and as free as wireless signals. The technologies Marinetti favored also acted as aesthetic "models" for his art; for example, in "The Electrical War (A Futurist Vision-Hypothesis)" (1911b/1971), Marinetti paid homage directly to the wireless and offered it as a model for his poetry.

The speed with which the wireless communicates (instantaneously) and the form an author's words take in telegraphic communication define Marinetti's "telegraphic lyricism" style. Such a style is "a swift, brutal, and immediate lyricism" (1913/1973, p. 104). Marinetti aimed to reduce words to the simplest, most efficient form possible. He wanted to recreate wireless transmissions in his poetry, privileging a telegraphic style. According to Campbell (2006), "Marinetti celebrates the ease with which the reproduced wireless practices will rid the poet of syntax" (p. 85). Various Futurist groups experimented with telegraphic styles, so Marinetti was not the sole originator. As White (1990) observed,

> Even granting Futurism an instrumental role in disseminating and popularizing telegraphic writing among the Orphists, the Expressionists, and the Vorticists, in the case of such a widespread modernist phenomenon one has to allow for the eventuality of polygenesis. Indeed, the possibility that the attractions of such a style had occurred almost simultaneously to various avant-garde writers in different countries is a strong one. (p. 147)

These European Futurist artists lived in industrial (izing) nations, so they and their aesthetics were products of their social context. The spirit of progress and modernization influenced this European "cult of the 'telegraphic,'" and "[i]ts mushrooming popularity owed much to two things: the discoveries of Guglielmo Marconi, and contemporary mythification of the Eiffel Tower" (White 1990, p. 147).

Although Marconi's invention was an important object for Marinetti, the promotion of progress underlies the goals set forth in his manifestos and poems. Besides his typographical long poem *Zang Tumb Tumb* (1914/1987), his early manifestos promoted this terse, telegraphic writing as an artist's duty, and embracing industrialization perpetuated human evolution. Marinetti can be used synonymously with Italian Futurist beginnings because Marinetti was the movement's leader,[6] and he is considered the most important (Blum 1996; Bondanella and Bondanella 1979; Hewitt 1993; Perloff 2003; Poggioli 1968; White 1990). From his first Futurist texts, readers are bombarded with industrial images. For instance, the first Futurist manifesto showed the movement's love of speed and violence in the prologue, where Marinetti (1909/1971) and his friends become re-energized after their car crashes into a river full of factory effluents (pp. 40–41). Such markers of industrialization pinpoint the correlation between art and the industrialized world's pro-technological values. "Growth," "progress," and "evolution" existed prior to Marinetti; after all, World's Fairs had been exciting the public with their technical, scientific, and industrial marvels for over half a century before Futurism. Marinetti, Marconi, and others were influenced by an age that celebrated humanity's techno-evolution. Machines were prostheses to humans, allowing them to accomplish "higher" forms of work. Increased production through mechanization meant humans could become superhuman workers. Marinetti also portrayed the machine as a model for human biology. To transform into a machine meant being able to cast off human "frailties" and become the most efficient specimen possible.

According to Marinetti, the Futurists, as super-efficient beings, "have already scattered treasures, a thousand treasures of force, love, courage, astuteness, and raw will power," yet "[o]ur hearts know no weariness because they are fed with fire, hatred, and speed!" (1909/1971, pp. 43–44). The image most appropriate for this transformation is to imagine a "radio-bomb." The radio signifies increased communication speed through telegraphic lyricism, and bombs reflect his incendiary rhetoric, which permeates Marinetti's poetry and prose.[7] An important aspect of his *parole in libertá* is "[l]ove of speed, abbreviation, and the summary. 'Quick, give me the whole thing in two words!'" (1913/1973, p. 98). As a "radio-bomb," Marinetti, who did not use this particular expression,[8] became a weapon launched from a radio tower to incite chaos. Although readers have to consider the figurative nature of poetry and prose, Marinetti's texts implied that he would have literally wanted to become an invective radio transmission exploding for millions within the reach of this new international communication technology. The wireless was a tool for the future, and "Marinetti singles out the invention of wireless telegraphy as one of the great milestones in civilization's progress towards the Futurist electric millennium of the twenty-first century" (White 1990, p. 148). As John White observed, the wireless excited the mass audience:

> News could be transmitted and received virtually instantaneously thousands of miles away. The telegraphic dissemination of the new word of science became the technological equivalent of the spreading of the Gospel, and writers treated the subject with an appropriate religiosity, often using the imagery of the biblical word. (p. 148)

From White's argument, the wireless's reception, fetishization, and subsequent reinscription can be said to have happened because the invention embodies the power of a divine miracle. Divinity had been replaced by technology, which Adams's (1900/1974) essay made clear.

Specifically, Futurism captured the culture's reliance and devotion to progress in works that ranged from manifestos and poems to sculpture and architecture. Each art genre—painting, literature, architecture, sculpture, etc.—had its own manifesto. Although Marinetti's movement was Italian, the value of mechanization, of progress through technological advancements, was international. Such sentiments ran through the European avant-garde circles and reflected the growing dependence on new sciences and technologies. In order to compete effectively in this hyper-industrial situation in which Marinetti and others found themselves, each nation had to acquire the proper technologies.

Marinetti's techno-dogma suggested that he attempted to help Italy see the need for a social push for modernization. During a visit to Venice, Marinetti and a few of his followers passed around leaflets chastising the "romantic" *passéisme* of the city. He wanted Italy to become a major industrial power, and he despised those who wished to reminisce about the past. The image is best captured by the comic strip artist André Warnod in *Comedia* from 1910:

> [T]he cartoonist provided comic 'before' and 'after' pictures…. Before: grotesque and flabby lovers smooching in the Piazza San Marco; after: a city of bridges dirigibles, smoke stacks, and electric lamps, their rays replacing those of the sun. To the left of center, we see a replica of the Eiffel Tower, as if to say Venice has now become Paris. (Perloff 2003, p. 104)

Paris, which had radio broadcasts beaming from the Eiffel Tower, was the pinnacle of modernization for Marinetti, a state Italy should mimic. Modernization aroused strong nationalist sentiments in Italy, and Marinetti himself conflated patriotism with a lust for mechanization. Because technology marked progress for Marinetti, anyone not consenting to the mechanical present (or future) "affirmed once again the ridiculous nullity of nostalgic memory, of myopic history and the dead past" (1911a/1971, p. 83). New technologies offered direction "from a new sun, which is certainly not the sun that caressed the placid backs of our grandfathers—those slow steps sagely measured to the lazy hours of provincial cities with their grassy cobblestones of silence" (1911a/1971, p. 83). Progress and progress reified through technological advancements were objects behind which pro-moderniza-tion forces rallied.[9]

2 The Effect of the Wireless and Other Technologies on Futurist Aesthetics

Just as Marconi and the popular press presented the wireless as another possible incarnation of modern progress, Marinetti unleashed his Futurist polemics urging further industrialization and, of course, mechanization. New technologies encour-aged progress because they extended the work a human could do: The automobile, assembly line, and wireless were all prostheses allowing humans faster movement, more efficient production, and wider communication possibilities. Prostheses like these helped create the aesthetic of dynamism: "The world the Futurists knew could be traversed… using such new means of transportation as the automobile, the high-speed train, and, for short runs, even the airplane" (Perloff 2003, p. xxxvii). In his texts, Marinetti imagined the human–machine possibilities of these new technolo-gies. Although one cannot deny the comfort of some technologies of leisure (cars, TVs, airplanes, etc.), the assembly line and factory itself are technologies of mechanization. Fordist/Taylorist management science sought to reduce the human (error) element in production by creating a situation (i.e., the assembly line) where workers' tasks could be reduced to simple repetition and little, if any, thought.[10] Such streamlining reflected dominant tropes of speed and efficiency.

Marinetti glorified technology by linking machine advancement to human progress—evolution. Current scholarship claims that Marinetti's aesthetics embody advancement through machinery. Although most scholars made the connection between technology and Marinetti's poetic style (Blum 1996; Butler 1994; Campbell 2006; Hewitt 1993; Perloff 2003; Taylor 1974), White (1990) specifically claimed that Marconi's wireless was a key influence on artistic telegraphy or, as he put it, "the cult of the 'telegraphic'" (p. 147). White also argued that Marinetti was not the only artist experimenting with such a style because "'telegraphic' writing was generally 'in the air' in European avant-garde circles" in the early twentieth century (p. 160). White credited Marconi's popularity and discoveries for inspiring "the public imagination and... hav[ing] a tangible influence on the quality of everyday modern life" (p. 148). In fact, Marinetti claimed that the wireless influenced his style after his time as a war correspondent during Italy's bombing of Tripoli (J. White, p. 161); however, White believes that account to be Marinetti's revisionist history (p. 162).

Christine Poggi argued that Marinetti's telegraphic *parole in libertá* writings "violate all the rules of proper telegram writing" (as cited in White 1990, p. 165). White noted that Poggi's argument for "the violation of the conventional rules for telegrams lies in Marinetti's use of diagrammatical pictures instead of any words" (p. 165). Focusing on Marinetti's violation of "proper" telegraphic form reduced his artistic contribution to *telegraphic lyricism* because "violation" assumes his style was somehow a miscommunication. Such a style was not meant to be of practical use in wireless or "wired" telegraphy. For example, Marinetti 's (1914/ 1987) famous bombardment poem *Zang Tumb Tumb* uses "**tataluuuntlin**" to describe the sound of a train going over an iron bridge and "**sssssssiii ssiissii ssiissssssiiii**" to describe the sound or whistle of a train's smoke stack (pp. 56, 57). Furthermore, the bolded text denotes loudness. Although a radio operator or telegrapher could easily record or decipher the letters above, Marinetti's onomatopoeic "words" were not meant to be communicated via Morse code. This style was meant to be read or heard (as it was in Marinetti's public performances). Marinetti's texts included such constructions as

poesia NASCERE

in order to signify a crescendo and to reinforce the idea that his poetry is born in this new age of bombs.[10] The crescendo, the gradual increasing of sound, mimics the gradual increasing loudness of a bomb falling to the ground, and the finale is an explosion.

The wireless influenced Marinetti's concept of *parole in libertá*, which offered readers a sense that contemporary communication technologies played a

Fig. 5.1 Image used by permission of Oxford University Press from *Literary Futurism: Aspects of the First Avant Garde* by White (1990) Image "8 Souls in a Bomb" p. 18

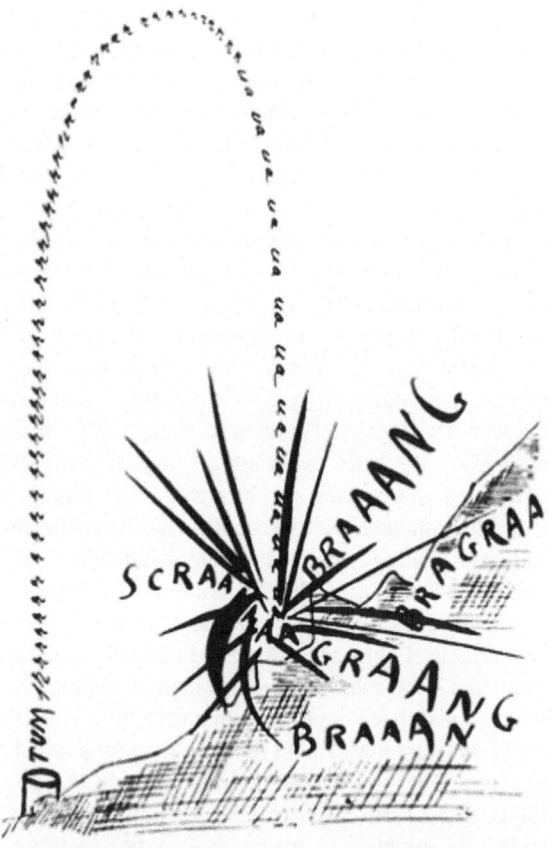

significant role in Futurist aesthetics. Marinetti embedded tropes of progress in his style. Marinetti's aesthetics derived cultural and artistic value from "progress" based on Industrialization's larger culture influences. Marinetti valued technology in a way that Latour's (1996) fictional sociologist in *Aramis, or the Love of Technology* would recognize as "love." Although Marinetti's response was more of an obsession than a practical technological design project, he reflected a larger cultural "love" of technology that allowed early twentieth-century inventions to become realized. Industrial cultures embraced new technologies and even felt in awe of their massive power, as Adams (1900/1974) did at the 1899 World's Fair. Marinetti felt humans should embrace technology in order to leap into the future. His devotion to ahistoricity was simultaneously his rejection of the past, a time, according to Marinetti, that was worthless.

   Marinetti also glorified war in his work. He enhanced his war imagery by using onomatopoeic and experimental typography. Besides a steadily increasing font size, Marinetti often "render[ed] dynamism typographically by using shaped writing to indicate *lines of movement*" (White 1990, p. 18). The cover of his 1919 novel *8 Souls in a Bomb* (Fig. 5.1) portrayed the sound of a bomb firing out of a cannon ("**TUM**"), going up into an arc ("**rrrrrr…**"), coming down onto a hillside ("**ua ua ua…**"), and exploding ("**Braaang Bragraa**"). This format was not supposed to be used as *sound* telegraphic style; instead, the typographical "image" conveys Marinetti's poetry as explosive. The eight souls of the title might have represented eight Futurist artists—F.T. Marinetti, Giacoma Balla, Antonio Sant'Elia, Gino Severini, Carlo Carrá, Umberto Boccioni, Luigi Russolo, and Andengo Soffici[11]—with the image evoked by the text's artistic representation of the Futurist polemical assault from the air and, possibly, the air *waves*. Although Boccioni, Russolo, and Sant'Elia died during (or from injuries sustained in) WWI, they were integral members pushing the boundaries of avant-garde art prior to WWI.

Those artists also celebrated movement, speed, and efficiency in their work, so the novel's cover represented the Futurist Movement's overall goal of exciting the imagination of the Italian masses through incendiary art. During a 1910 performance in Venice, Marinetti and his cohorts dropped thousands of leaflets with an anti-*passéisme* manifesto directed at the Venetian crowd because they were not embracing modernization enough. The manifesto titled "Marinetti's Futurist Speech to the Venetians" ended with on a scolding tone:

> Shame on you! Shame on you! And you throw yourselves one on top of another like bags of sand to make an earthworks on the border, while we prepare a great strong, industrial, commercial, and military Venice on the Adriatic Sea, that great Italian Lake! (Marinetti 1910/1971, p. 58).

As Campbell (2006) mentioned, "The vehemence with which the Futurists supported Italian intervention in 1915 and their success in registering its movements in 'art'…confirm that the wireless's command to speed enrolls bodies in the service of war" (p. 89). Later in the 1920 and 1930s, Marinetti would douse his audience with his polemical Variety Theatre and radio broadcasts, but the pre-WWI violent rhetoric existed on the page in Marinetti's earlier works, much of which glorified bombardment and communication as ways to attack. Other technologies contributed to Marinetti's violent poetry, but his "wireless imagination" concept created an ideal freedom inextricably linked to mechanical values and tropes of progress. After all, the mechanical wireless produced signals that flew back and forth between stations.

One of Marinetti's most celebratory works specifically about the wireless is his poem-manifesto[12] "Electrical War (A Futurist Vision-Hypothesis)" (1911b/1971), which showed how the wireless was reconceived through bombardment imagery to become both a muse of sorts and a means to promote Italian nationalism. This pre-WWI piece heralded a new *panitalianismo*, which was to be carried out in large part by new technologies. "Electrical War" opens with the following introduction: "Oh! how I envy the men who will be born into the next century on my beautiful peninsula when it is wholly vivified, shaken and bridled by the new electric forces!" (p. 104).[13] The wireless acted as an important controlling mechanism for Marinetti's imagined, future Italy "transformed by man's genius into many millions of Kilowatts" (1911b/1971, p. 104). In the vision, Marinetti denounced the romanticist poets of the past for their love of the sea and claimed humans had found a way to make the sea *labor* for Italy "with all its diligent, raging storms to set in ceaseless motion numberless iron pontoons that energize two million dynamos scattered along the beaches and in a thousand working gulfs" (1911b/ 1971, p. 104). This energy, "needing no wires" in Marinetti's vision, was to be "controlled from keyboards with a fertilizing abundance that throbs beneath the fingers of the engineers" (1911b/1971, p. 104). Instead of having a world created by natural forces, Marinetti's vision promoted a hyper-industrial science fiction reality where radio-type engineers "sit before switchboards, with dials to right and left, keyboards, regulators, and commutators, and everywhere the splendid flash of polished levers" (1911b/1971, p. 104–105). While this "control room" reflects a science fiction spaceship for readers today, Marinetti's description came from a radio room. The vision's engineers "have finally won the joy of living between iron walls," and, because "[t]hey are finally free of wood and its lesson of weakness," mechanization forged a new reality based on the strength of new materials (1911b/1971, p. 105). Marinetti privileged machines for their power and denigrated the natural, which he saw as weakness needing to be conquered. Not surprisingly, this vision came from his novel *War, the World's only Hygiene* (1915) where he glorified the cleansing aspects of destruction—destruction of the past carried out by new technologies.

Airplanes and the wireless assisted the vision's engineers who "regulate the lightning *speed* of the seed-scattering trains that two or three times a year cross the lowlands for basic sowing" (1911b/1971, p. 105, emphasis added). Airplanes had yet to have wireless capabilities in 1911, but popular press articles had already prophesized remote control possibilities—mechanical manipulation via radio signals—so a radio in an airplane would not have seemed too far fetched. One journalist even speculated that remote control torpedoes would eventually be used during naval battles (Moffett 1899, p. 17). While popular press reports speculated on non-organic wireless and general electrical possibilities, Marinetti had a more fantastic, organic vision where electricity not only planted seeds but helped them grow. Marinetti envisioned crops and orchards with "numberless lightning rods" placed in the ground to "tickle the turgid bellies of storm clouds" in order to excite "the roots of the plants" (1911b/1971, p. 105). This electro-fertilization helped grow bigger, stronger crops faster, a system, although implausible, that was sup-posed to be far more efficient than simply organic means. Electricity, which appeared to follow the idea that ether was a conductor, was no longer wasted: "All the atmospheric electricity hanging over us, all the incalculable electricity of the earth, is finally harnessed" (1911b/1971, p. 105). Plants also grew "with lightning speed" when stimulated by "artificial electricity at high tension" (1911b/1971, p. 105). The wireless made this electro-agricultural vision possible because invisible electric signals—with no physical connection—had no limits for Mari-netti. Wireless signals penetrated walls, hills, and other barriers, so electricity "will penetrate every muscle, artery, and nerve of the [Italian] peninsula" (1911b/ 1971, p. 104).

The Futurist vision assumed that "the discovery of the wireless telegraph far in the past" helped bring about this techno-utopian Italy (1911b/1971, p. 106). Marinetti's glimpse of the future reflected Marconi and the popular press's topos that the wireless's potential was beyond expectations, and that the future would fully recognize the wireless and other technologies' capabilities. The wireless and the fact that "the use of dielectrics increases every day" (1911b/1971, p. 106) built Marinetti's vision of a hyper-industrial and hyper-efficient new world. Just as Marconi and supportive popular press writers rhetorically constructed the wireless as a commercial success, Marinetti infused his poem-manifesto with labor images, which demonstrates both the popular press and Marinetti promoted industry. In Marinetti's utopian world, physical occupations were done by machines: "Ended now the need for wearisome and debasing labors. Intelligence finally reigns everywhere. Muscular work ceases to be servile" (1911b/1971, p. 106). Marinetti also promotes a Libertarian or proto-Objectivist theory[14] because *laissez faire* economic principles appeared to give humans massive surpluses. No hunger or poverty existed in this vision because "[t]he financial question [has been] reduced to a simple matter of accounting" and society allowed "[f]reedom for all to make money" (1911b/1971, p. 106).

Because of these new commercial successes, Marinetti created a new ideal for human intellect and prosperity. Basically, he wanted machines to (do the) labor, so that these new humans "can perfect their lives in numberless antagonistic exertions" (1911b/1971, p. 106). The new humans no longer walked on earth—too inefficient; instead, they flew in personal monoplanes. Physical exertion "now [has] only three goals: hygiene, pleasure, and struggle" (1911b/1971, p. 106). Even today pro-industrialists promote technology as saving human time. Machines are seen as more efficient, which is why athletes and coaches often describe a person's body through machine metaphors. A machine does not get sick or perform inconsistently in such comparisons. If athletes have automatic, mechanical physical responses at game time, they should win. Marinetti envisioned such cyborg-like creatures who could now be more efficient because "[e]yes and other human organs are no longer simple sensory receptors, but true accumulators of electric energy" (1911b/1971, p. 106).[15] In other words, Marinetti celebrated as human evolution the mechanical and human worlds becoming entwined after technological efficiency had replaced human labor.

For Marinetti, advancement broke from the "old ways," so techno-celebrations fit Futurist art well. Because technology is valued for commercial reasons, its promoters (as I have shown in chapters three and four) emphasize commercial potentials. Marinetti was no exception. The world he imagined included "[t]wenty five great powers... fighting over markets of a superabundant industrial production" (1911b/1971, p. 107). Marinetti envisioned that "the first electric war" would not use "more of those old explosives" but would harness the energy of the atmosphere and attack using wireless means—signals as bombs (1911b/1971, p. 107). In this poem-manifesto, the wireless was a central technology that helped to usher in this new utopian (albeit violent) hyper-industrial world. The world powers were to fight over resources in cleansing wars. Marinetti felt war was the world's only hygiene, for it would remove the past and the inefficient elements that stood in the way of progress.

To indicate progress and a break from the past, Marinetti's vision offered a horrific image: "The sick and weak, crushed, crumbled, pulverized by the vehement wheels of intense civilization. The green beards of provincial back alleys will be shaved clean by the cruel razors of speed" (1911b/1971, p. 108).

Before this new world could begin, Marinetti claimed Italians must "kill" the past. In "The Founding and Manifesto of Futurism" (1909/1971) which established Marinetti's movement, he rejected the past and was born anew. He advanced his attack further in "The Electrical War (A Futurist Vision-Hypothesis)" (1911b/1971) by denouncing *passéisme*. Marinetti ridiculed lovers of the past and those who loved Italy specifically for its past. To "correct" this seemingly misguided love he offered the following plan:

> After having insulted every stranger who adores our Italian past and despises us as singers or serenades, as ciceroni or beggars, we have asked them to admire us as the most gifted race on earth. Thanks to us, Italy will cease to be the love-room of the cosmopolitan world (1911b/1971, p. 108).

Marinetti wanted Italy to evolve into this utopian vision through technology.

3   A Love of Technology

Marinetti's vision captured a love of technology and dismissed any remotely nostalgic aesthetic. While *telegraphic lyricism* was a common component of the European avant-garde circa 1910 (Perloff 2003; White 1990), the techno-fetishization apparent in Marinetti's art was predominantly Futurist. White (1990) argued that Marinetti's *parole in libertá* style could be "interpreted as a part of a process of liberation from the hold of the past which is an inevitable feature of the modern world" (p. 164). J. White also claimed that the Futurists' love of brevity "is more than just a virtuous economy reflecting the pace of modern life"; instead, "it is in some way the key to greater truths and thus renders most passéist writing not only anachronistic but also superficial" (p. 164). Marinetti's style reflected the industrial world's cult of efficiency more than any link to actual telegraphic writing practices. He envisioned a smaller world "brought about by the great discoveries of science" (as cited in Perloff 2003, p. 57), and he promoted these new discoveries in his poetry and prose. His manifestos and telegraphic writing pointed to the technological and scientific undercurrent of his time period.

Marinetti positioned the wireless as the new, efficient communication technology helping to make the world smaller. He also portrayed other technologies—often in the image of a generic machine—as indicative of modern progress. Marinetti embedded tropes of progress in his texts, which serve as sites where Marinetti's early Futurist stance can be read as propaganda for a future that was to glorify technological advancement as reified *progress*: "Marinetti's argument [for speed and efficiency] is reductive enough to make for compelling propaganda" (Perloff 2003, p. 57). Although such a discussion could highlight Marinetti's proto-fascist leanings, readers should avoid reading fascism into Marinetti's early works because that would risk being reductive as Perloff warned (p. xxix).[16] Marinetti's glorification of violence, war, and industrialization was not uniquely fascist. After all, quite a few parallels exist between the Futurist Manifestos, Marconi's presentations, and the popular press's re-presentations. To read his works as fascistic ignores the richer discussion that argues fascist principles are systemic features of industrial societies enamored with new technologies. Even "democracies" such as the United States or Britain had more of an oligarchy controlling resources and production. In order for industries to profit, they needed highly efficient and dependable labor.

The new materials (and material conditions) of the early twentieth century required new ways of organizing resources. Taylor (1911/1967) advocated his management science theories to help make factories more efficient. Marinetti, on the other hand, promoted efficiency as human virtue, one developed by privileging mechanical qualities over human frailty. To label such an aesthetic quality or industrial practicality as *fascist*, ignores the ideological link to all industrial economies—fascist, communist, capitalist, or "mixed" systems. The wireless and Futurism are both products of modernity, and two important events in 1909 point to the significance of science and technology in popular and literary consciousness: Marconi won the Nobel Prize for Physics, and Marinetti published his first Futurist manifesto. Marconi's Nobel Prize did not introduce him to the world but helped *certify* his status among the scientific community. In contrast, Marinetti, who "was a mediocre late Symbolist" until his seminal manifesto (Perloff 2003, p. 84), exploded onto the new avant-garde literary scene.

Marinetti's manifesto and Marconi's Nobel Prize both reflected the industrialized world's positive attitude toward progress. Just as Marconi and the popular press presented the wireless as a marker of "good" civilization, Marinetti used the same tropes of progress to praise technologies: speed, efficiency, evolution, and ahistoricity. Marinetti's works show technology as a new way to liberation—freeing humans from their servile manual labor. Such a situation distinctly marked human progression away from the past. For Marinetti, industrialization would usher in a new Italy, so past attributes or history itself was irrelevant.

4 Ahistoricity in Marinetti's Aesthetics

Marinetti's most obvious trope of progress extended from his disdain for the past and promotion of ahistoricity. In Marinetti's "Founding and Manifesto of Futurism" (1909/1971), Futurism was born in an allegorical "escape" from a backward, non-industrial countryside: After an evening contemplating where art should go, Marinetti and his friends "trampled our atavistic ennui" and roared away in an automobile (1909/1971, p. 39). Marinetti signaled their departure from the past by shouting "Mythology and the Mystic Ideal are defeated at last.... We must shake the gates of life, test the bolts and hinges" (1909/1971, p. 39–40). After a series of quick twists and turns while roaring down the highway, Marinetti crashed into a ditch of "[f]air factory drain" (1909/1971, p. 40). The polluted river revitalized Marinetti: "And so, faces smeared with good factory muck—plastered with metallic waste, with senseless sweat, with celestial soot—we, bruised, our arms in slings, but unafraid, declared our high intentions to all the *living* of the earth" (1909/1971, p. 41). Thus begins the first avant-garde manifesto.

As the name implies, Futurism looked forward, and Marinetti clearly showed his loathing for *passéisme* in his early works. The eighth point of the first manifesto claimed, "We stand on the last promontory of the centuries!... Why should we look back, when what we want is to break down the mysterious doors of the Impossible? Time and Space died yesterday" (1909/1971, p. 41). Marinetti also claimed to want to "destroy the museums, libraries, academies of every kind" and "fight moralism, feminism, every opportunistic or utilitarian cowardice" (1909/1971, p. 42). Only the new was to be admired in both art and technology. Marinetti attacked *passéisme* by

> establish[ing] Futurism, because we want to free this land from its smelly gangrene of professors, archaeologists, ciceroni and antiquarians. For too long has Italy been a dealer in second-hand clothes. We mean to free her from the numberless museums that cover her like so many graveyards. (1909/1971, p. 42)

He rails against museums, claiming they were cemeteries, and "[a]dmiring an old picture is the same as pouring our sensibility into a funerary urn instead of hurtling it far off, in violent spasms of action and creation" (1909/1971, p. 42). Being linked to the past stopped "action and creation," signifying death or halting progress.

Although denying the past and stressing progress(ion) were avant-garde characteristics, these were also the tropes of business. Taylor (1911/1967) asserted that "great gain, both to employers and employés," will come "from the substitution of *scientific* rule-of-thumb methods in even the smallest details of the work of every trade" (p. 24, emphasis added). Taylor's text was also a manifesto of sorts. He, as did Marinetti, placed the onus on the individual to eliminate sloth and inefficiency. Inefficient workers were guilty of dereliction of duty. Taylor argued that "[t]he enormous saving of time and therefore increase in output... can be fully realized only after one has personally seen the improvement" of Taylor's scientific application (p. 24). One major goal of Taylorism was efficiency from "[t]he general adoption of scientific management" to achieve "the increase, both in the necessities and luxuries of life, which becomes available for the whole country" (p. 142). Also, another goal would be "the elimination of almost all causes for dispute and disagreement between [management and workmen]" (p. 142). According to Taylor, applying scientific principles to the factory led to efficiency and benefits all because science touched all workers and helped produce goods efficiently. Likewise, Marinetti found a similar attitudinal change because "Futurism is grounded in the great discoveries of *science*" (1913/1973, p. 96, italics mine). For both men, science was a good organizational strategy for either factories or art.

Although Taylor (1911/1967) directly addressed managers and workers, the results of his system were to be a benefit to all industrialized nations: "Is it not the duty of those who are acquainted with these facts, to exert themselves to make the whole community realize this [study of scientific management's] importance" (p. 144). Taylor's text had a certain avant-garde quality inherent in its hyper-industrial fervor, but it was hardly the polemic of Marinetti's art. Taylorism (and its famous put-in-practice system, Fordism) adhered to early twentieth-century ideology—speed, efficiency, evolution, and ahistoricity; it was, in fact, a product of the time period. Besides promoting progress, Taylor chastised his inefficient audience as morally defunct, thus satisfying Renato Poggioli's (1968) definition of the avant-garde: "Ideology, therefore, is always a social phenomenon. In the case of the avant-garde, it is an argument of self-assertion or self-defense used by a society in the strict sense against society in the larger sense" (p. 4). Taylor's manifesto was a product of the time, and a rubric for adhering to the value of efficiency. Factories needed only follow the principles Taylor put forth, and they would assert their productive dominance in the market. Taylor advocated his "primer's" value for the larger society would be modernization, the same argument Marinetti made when he advocated "killing" any connection to the past would help Italian modernization.

I am not arguing, however, that Taylor shared Marinetti's ahistorical stance against cultural markers and artifacts. While Marinetti's work was prone to violence, exaggeration, and performance, Taylor appeared more practical, systematic, and *industrious*. Taylor privileged the worker and management's role in maintaining a well-organized firm. Ultimately, he does not fit Poggioli's (1968) definition of an avant-garde artist because his work was not absorbed into "the demagogic moment," which Poggioli argued fueled the "[avant-garde's] tendency toward self-advertisement, propaganda, and proselytizing" (p. 34). Although Taylor's lack of gross exaggeration and ferocious polemical stances mitigated his avant-garde status, his importance for gauging industrial practice is without question: His text existed as a heuristic for industrial progress. Simply put, *progression* toward increased production and profits mirrored part of Marconi and the popular press's rhetoric of technology. The wireless, besides often being "praised" for its potential, reflected human advancement and commercial/industrial success. In this historical moment, Taylor (1911/1967) claimed,

> our larger wastes of human effort, which go on every day through such of our acts as are blundering, ill-directed, or inefficient, and which Mr. [Theodore] Roosevelt refers to as a lack of 'national efficiency,' are less visible, less tangible, and are vaguely appreciated (p. 5).

Marinetti provided a symbolic transition. Instead of looking to the past's supposed "summit" or "fullness of time" as a goal for avoiding "a fatal infelicitous fall back to barbarism" (Poggioli 1968, pp. 72–73), Futurism experimented with the new.

These new experiments, although brash and violent, glorified new technologies that were unconsciously accepted by industrial cultures. Of course, a cultural studies lens cannot identify all values a society in a particular time period had. However, prevailing values appear during cultural studies research. Technologies

expose ideological tenets because they do not come to be without group acceptance. Because researchers have the benefit (or burden) of historical hindsight, we
know that the wireless extended the reach of communication—it was heralded as a
genius product of modernity. Likewise, automobiles became accepted as beneficial
technologies, "liberating" individuals in industrialized nations because of their
potential for allowing greater mobility. Today, wireless transmissions, automobiles, and other technical objects are more than just tools; these technologies are
prostheses for human activity in industrial, hyper-technological societies.

Technologies do not have to be accepted universally in order to become realized. We cannot claim every member of a society uses such technologies only that
they are popularly seen as efficient, necessary products we cannot live without. As
long as large enough groups accept a certain technology, these tools will be seen as
useful and, therefore, be realized. In fact, these technologies (and "universal"
technologies like computers, PDAs, or mobile phones) can really only be said to be
prostheses for middle, working, and wealthy classes. Claiming "everyone has a
mobile phone" marks the chauvinistic impulse in dominant society to ignore the
material conditions of poorer groups. Such chauvinism appeared in Futurism
specifically and avant-gardism generally, which "is by nature solitary and aristocratic" (Bontempelli as cited in Poggioli 1968, p. 39). Marinetti claimed "I do
not care for the comprehension of the multitude," and that poetry, avant-garde or
traditional also "requires a special speaker if it is to be understood" (1913/1973,
p. 106). Likewise, because new technologies provide markers for civilization, a
citizen must acquire the appropriate artifacts to be in accord with the well-to-do
members. Consumerism allows individuals of any background to "buy into" the
aristocratic image. An aristocratic technology such as the wireless held a certain
regal aura because of how favorable relevant social groups rhetorically constructed
it. Marconi and the popular press documented when royalty and national leaders
used the wireless, constructing it as an aristocratic or "elevated" technology.
However, for a small fee, any individual could send a wireless telegram (Baker
1902, p. 12), allowing him or her access to an aspect of an aristocratic lifestyle.

Marinetti addressed these aristocratic desires through tropes of efficiency and
the new efficient, machine-like aristocrat. But, because of Marinetti's disdain for
tradition, the old aristocrat (the noble or royal) could not serve as a model.
Although Marinetti came from a wealthy family (Bondanella and Bondenella
1979, p. 315) and, as Bontempelli points out, "[the avant-garde] loves the initiated
and the ivory tower" (as cited in Poggioli 1968, p. 39), the new aristocrat must
embrace modernization. Marinetti praised the industrial (ized) aristocracy over the
royal by promoting aesthetic goals against ornamentation: "A *modern* aesthetic
most responsive to *utility* has no need of royal palaces with domineering lines and
granite foundations that loom massively *out of the past* over the little medieval
towns, confused welters of wretched dog kennels" (1911a/1971, p. 80–81,
emphasis added). Instead, the "definitive Futurist aesthetic" included "great
locomotives, twisting tunnels, armored cars, torpedo boats, monoplanes, and
racing cars" (Marinetti 1911a/1971, p. 81). The new "modern phenomena"—
technologies forged from modern ideals—succeed in "hav[ing] reduced to

uselessness the great, decorative, imperishable buildings that once expressed kingly authority, theocracy, and mysticism" (Marinetti 1911a/1971, p. 80). In order to progress and fully enjoy modern life, such as "the speed of international communications," modern comforts from "well-ventilated apartment blocks" to "perfect *chambers de toilette*" were required (Marinetti 1911a/1971, p. 80). These technologies, these *phenomena*, signaled progress to Marinetti.

5 Progress Tropes in Marinetti's Art

Marinetti did not stop by listing the important new technologies of the time period; he also celebrated "progress" as an abstract goal useful for his new aesthetic and ideal for life. Humans should strive for progress and hold dearly to the concept as a defining goal. Marinetti's rhetoric of technology favoring progress matched Marconi and the popular press's rhetoric because all three discourses advocate technological advancement as inherently *progressive*. Progress was the code by which society should live: "Put your trust in Progress, which is always right even when it is wrong, because it is movement, life, struggle, hope" (Marinetti 1911a/ 1971, p. 82). Marinetti's own capitalization of the word "Progress" reinforced the concept's importance for the new modern world. Progress and constant motion (dynamism) were both elements of Marinetti's aesthetics reinscribed into Futurist art. Marinetti's art—pushing dynamism as one goal—privileged the struggle or act of creation over a finished product: "The frame of a house in construction sym- bolizes our burning passion for the coming-into-being of things. Things already built and finished, bivouacs of cowardice and sleep, disgust us!" (1911a/1971, p. 82). Futurist sculpture, painting, and even architecture have widely been ana- lyzed for their adherence to dynamism (Butler 1994; Kirby 1971; Perloff 2003; Poggioli 1968; Rainey 1998; Rye 1972). Only through dynamism in art and life can one find beauty: "Except in struggle, there is no more beauty. No work without aggressive character can be a masterpiece" (Marinetti 1909/1971, p. 41). Marinetti saw progress reified through construction "according to the ever-changing moods of the winds" (1911a/1971, p. 82).

This change supported Marinetti's view that "the world's magnificence has been enriched by a new beauty, the beauty of speed" (1909/1971, p. 41). According to Marinetti, speed is limited when humans cling to the past. Italy's past is a "heavy burden... that weighs down our swift and warlike vessel" (1911a/ 1971, p. 83). That burden of the past worked against modernization, and Mari- netti's claim that "we Italian Futurists have no desire to see Italy left in an inferior state" (1911a/1971, p. 83) reflected the rhetoric behind promoting technology as a way out of "barbarism." Although *barbarism* is an exaggerated term, the concept is what the time period read as evolution's antithesis (c.f., Childs 2000). Just as today's educational push for more math and science skills is argued as *crucial* to a country's economic prosperity, lack of industry in the early twentieth century marked a nation as backward. A January 1902 article from *The North American*

Review claimed that "America's position in the world of science is inferior," and that Americans—even with "the stimulating examples of Edison, Tesla, [and] Elihu Thomson"—have been slow to pursue scientific enterprise (Snyder 1902, p. 59).[17] In order to remedy this "dire" situation, the author claimed Americans should create an equivalent of England's Royal Institution; however, for this situation to yield results "would require... those who are broadly interested in scientific progress, and [those] who have a desire to keep abreast of the swiftly advancing knowledge of the day" (Snyder 1902, p. 59). Science and technology advance rapidly, and both Marinetti and the article promoted a rhetoric that identified speed and advancement as a cultural value.

In Marinetti's words, Futurism "create[d] the new aesthetic of speed," which "notably diminished the concept of time"; before long "[w]e will arrive at the abolition of the year, the day, and the hour" (1911a/1971, p. 81). All that remains is the *quickness* of a moment. But Marinetti did not argue for "speed" in his art just by presenting images of racing cars and fast-moving ocean liners; he linked the reader's imagination to the speed of telegraphy by claiming "the analogical foundation of life" required quick communication "with the same economical *speed* that the telegraph imposed on reporters and war correspondents in their swift reportings" (1913/1973, p. 98). This new aesthetic reaffirmed what Marinetti argued was the relationship between poet, audience, and industrialization: "This urgent laconism answers not only to the laws of speed that govern us but also to the rapport of centuries between poet and audience" (1913/1973, p. 98).[18] Within this new aesthetic, Marinetti described certain tenets or textual properties that favored speed, such as Futurism's "[d]read of slowness, pettiness, analysis, and detailed explanations" (1913/1973, pp. 97–98).

Speed and efficiency were major modernist tropes and topoi for Marconi and the popular press, and Marinetti's embedding of efficiency into his aesthetic and his glorification of it signaled another rhetorical reconstruction of industrial values. For Marinetti, technologies did not just *mark* progress, they helped humans *achieve* progress. New technological discoveries "have a decisive influence on [users'] psyches" (Marinetti 1913/1973, p. 96) because a human becomes intertwined with the machine's efficiency; after all, Marinetti felt technologies acted as prostheses, creating "[m]an multiplied by the machine" (1913/1973, p. 97). This "[n]ew mechanical sense" Marinetti attributed to humans created a *better* worker through "a fusion of instinct with the efficiency of motors and conquered forces" (1913/1973, p. 97). The height of efficiency for Marinetti's new human would come from what we today know as cyborgs—*cybernetic organisms* that are part human and part machine. While "preparing the ubiquity of multiplied man", Marinetti sets forth a concept that violently dehumanizes "regular" humans (1911a/1971, p. 81). In order to be efficient, one must use the new technologies simultaneously; one must multi-task. Others have pointed out that Futurism had Nietzschean roots (Bondanella and Bondanella 1979; Childs 2000; Perloff 2003; White 1990), and Marinetti's concept of the "multiplied man" was nothing more than an *übermensch* made "super" through technological prostheses. Workers extend their productive capacity by replicating mechanical responses, and these "most gifted people" are also "the most elastic [and] quick" (Marinetti 1911a/ 1971, p. 83). Such malleability—and speed from having a worker mimic a machine—show dehumanization: An individual has been reduced to a machine part, a cog, perhaps. This individual is multiplied throughout a factory when the workers cling to (or are made to embrace) scientifically managed labor.

Scientifically or technologically reproduced "humans" appear in popular visions of cyborgs today,[19] but an unfavorable contemporary portrayal of scientifically managed workers appeared in D. H. Lawrence's *Women in Love*. The novel mainly follows the relationships of two sisters, Gudrun and Ursula, who fall in (and out) of love. Ursula and Rupert are the novel's eventual happy couple who stand by one another. In contrast, Gerald and Gudrun have difficulties connecting to each other, and their relationship disintegrates. But before the demise, Lawrence (1920/1995) described Gerald's new management techniques for his mines as techniques scientifically "formulated" to produce efficient results. Gerald believes, as the boss "above" the workers, he is "giving them what they wanted" because his firm is part of the "great and perfect system that subjected life to pure mathematical principles" (Lawrence 1920/1995, p. 231), which is the Futurist credo. Gerald's miners did not like his new management practices at first, but they eventually "submitted to it all" (Lawrence 1920/1995, p. 230) because this was the future—human evolution through mechanized frameworks:

> There was a new world order, a new order, strict, terrible, inhuman, but satisfying in its very destructiveness. The men were satisfied to belong to the great and wonderful machine, even whilst it destroyed them. It was what they wanted, it was the highest that man had produced, the most wonderful and *superhuman*. (Lawrence 1920/1995, p. 231, emphasis added)

Gerald runs the company "on the most accurate and delicate scientific method"; thus, "the miners were reduced to mere mechanical instruments" (Lawrence 1920/ 1995, p. 230). Of course, this Taylorist/Fordist impulse for ultra-efficiency was part of the rise of the industrial West. Lawrence appeared to be aware of America's dominance (or coming dominance in industrialization) because he wrote that the "[n]ew machinery was brought from America, such as the miners had never seen before, great *iron men*, as the cutting machines were called, and unusual appliances" (1920/1995, p. 230, emphasis added). Even the machines had nicknames to reinforce their "human" qualities.

Gerald became the workers' god: "He was the God of the machine. [The miners] made way for his motor-car automatically, slowly" (Lawrence 1920/1995, p. 223). Gerald, literally, kept the gears turning; he would dispose of non-critical, inefficient parts (workers) as he needed. An apotheosis of mechanization is apparent in Gerald's "incarnation of his power, a great and perfect machine, a system, an activity of pure order, pure mechanical repetition, repetition ad infinitum, hence eternal and infinite" (Lawrence 1920/1995, p. 228). Besides reinforcing the repetitive nature of the industrial gear, Lawrence exposed science or, in this case, mechanization begetting mechanization over and over again. This circular framework of Gerald's mind—and heart—foreshadows the dead end where he arrives at the novel's conclusion. The mechanization loop is not a loop an industrialist can escape. It is a self-contained view of production and progress that does not allow for new phenomena to develop. Humans will only "evolve" by going round and round on the merry-go-round of circular technological advancement, *ad infinitum*. Gerald is the god in his own mind because he adheres so vehemently to the cult of efficiency that permeates industrialization. By putting humans into this framework, their worth directly relates to the work they produce: "The sufferings and feelings of individuals did not matter in the least....What mattered was the pure instrumentality of the individual" (Lawrence 1920/1995, p. 223).

Unfortunately, this system sets humans up for failure because the loop continually pushes them to work faster and faster, which is "terrible and heartbreaking in its mechanicalness" (Lawrence 1920/1995, p. 230). Therefore, Gerald is not inherently corrupt and destined to breakdown as a machine would, but he is a victim of his own dogma, an ideology the industrialized world embraces along with the cult of efficiency. Gerald, entwined in the pursuit of hyper-mechanization, absorbed the value that he must keep moving to be worth anything, and he is overcome with fear when he succeeds at his "exalted activity" of running the company down to a science: "And once or twice lately, when he was alone in the evening and had nothing to do, he suddenly stood up in terror" and went to the mirror to look at himself and saw "the darkness in [his eyes]" (Lawrence 1920/ 1995, p. 232). For Gerald was "afraid that one day he would break down and be a purely meaningless babble lapping round a darkness" (Lawrence 1920/1995, p. 232). This terror Gerald feels relates to his inability to connect to women other than for surface affection, and it foreshadows his breakdown on the mountain after he realizes Gudrun will never be his wife and that he became a monster by almost killing her.

Such mechanical (in) humanity is ridiculous to Lawrence (1914/1979), who, in a letter to Arthur McLeod, claimed that the Futurists were rather sophomoric in their art. Although he was attracted to the Futurists' attempts "of the purging of the old forms and sentimentalities," which he appreciated "for its saying—enough of this sickly cant, let us be honest and stick by what is in us" (p. 180), he rejected them as "very young, infantile, college student[s] and medical-student[s] at [their] most blatant" (p. 180). Lawrence viewed Futurism as a young movement, but he did not see it maturing; instead, "[t]hey will progress down the purely male or intellectual or scientific line. They will even use their intuition for intellectual and scientific purpose. The one thing about their art is that it *isn't* art, but ultra scientific attempts to make diagrams of certain physic or mental states" (p. 181). The Futurists' "revolt against beastly sentiment and slavish adherence to tradition and the dead mind" were aspects Lawrence praised in the Futurists, but he "[doesn't] agree with them as to the cure and the escape"—the obvious militaristic, mechanistic nature of the movement (p. 181). From this letter, Lawrence showed he was clearly distrustful of scientific laws being applied to human consciousness, for this was the downfall of Gerald in *Women in Love* and a critique of the over-mechanization of thought: inhuman desire to act like a machine.

Such action led to Marinetti's most dehumanizing vision of a worker dying for the "noble" cause of construction. Marinetti valued efficient human production to such an extent that job sites should hear "from time to time—*yes, let it happen*—the harrowing cry and heavy thud of a fallen construction worker," causing Marinetti to exclaim "great drop of blood on the pavement!" (1911a/1971, pp. 81–82, emphasis added). Workers were valued for efficiency: Their humanity had been reduced to what their labor produced, and they were expected to give even the *ultimate* sacrifice if need be. This sacrifice contributed to Marinetti's rhetoric of technology because he valued struggles to create new machines. Of course, his manifestos were exaggerations of hyper-industrialization, but they mirrored Marconi and the popular press's topoi of presenting technologies (and modernization in general) related to tropes of progress.

Progress permeated the technological and scientific literature I have examined from the early twentieth century. Marinetti's early Futurist texts demonstrated how one important historical avant-garde managed to reinscribe dominant tropes of progress into the movement's aesthetics. Marinetti captured the essence of progress in his texts by advocating the same values and practices of the larger culture; specifically, he reflected industrial society's push for modernization by glorifying contemporary technologies and arguing for their future potential. Because technologies marked an advanced civilization, the Futurists saw new technologies as a way to bring Italy further progress and become a world industrial leader. Although such advocacy was an exaggeration of societal values and goals, Futurism embodied the ideologies favoring science and technology and portrayed new advancements as expanding human capacity. Technology thrusted humans toward the future, and, according to Marinetti, "[a]ll... hope should be in the Future" (1911a/1971, p. 82).

References

Adams, H. (1974). The dynamo and the virgin. In E. Samuels (Ed.), *The education of Henry Adams*. Boston: Houghton. (Original work published in 1900).

Baker, R. S. (1902). Marconi's achievement: Telegraphing across the ocean without wires. *McClure's Magazine, 18*(4), 4–12.

Blum, C. S. (1996). *The other modernism: F. T. Marinetti's futurist fiction of power*. Berkeley: University of California Press.

Bondenella, P., & Bondenella, J. C. (1979). *Dictionary of Italian literature*. Westport: Greenwood Press.

Butler, C. (1994). *Early modernism: Literature, music and painting in Europe, 1900–1916*. New York: Clarendon Press.

Campbell, T. (2006). *Wireless writing in the age of Marconi (electronic mediations)*. Minneapolis: University of Minnesota Press.

Childs, P. (2000). *Modernism*. London: Routledge.

Ford, H. (1999). Making history. In R. Rhodes (Ed.), *Visions of technology: A century of vital debate about machines, systems and the human world* (p. 61). New York: Touchstone. (Original work published in 1916).

Hewitt, A. (1993). *Fascist modernism: Aesthetics, politics, and the avant-garde.* Stanford: Stanford University Press.

Kirby, M. (1971). *Futurist performance.* New York: PAJ Publications.

Latour, B. (1996). *Aramis, or the love of technology.* Cambridge: Harvard University Press. (C. Porter, Trans.).

Lawrence, D. H. (1914/1979, June 2). Letter 731. In G. J. Zytaruk and J. T. Boulton, (Eds.), *The letters of D. H. Lawrence* (Vol. 2, pp. 180–182). Cambridge: Cambridge University Press. (Letter to Arthur McLeod).

Lawrence, D. H. (1995). *Women in love.* London: Penguin. (Original work published in 1920).

Marinetti, F. T., (1909/1971). The founding and manifesto of Futurism. In R. W. Flint, (Ed.), *Marinetti: Selected writings* (pp. 39–44). New York: Farrar, Straus and Giroux. (R. W. Flint & A. A. Coppotelli, Trans.).

Marinetti, F. T. (1910/1971). Marinetti's futurist speech to the Venetians. In R. W. Flint, (Ed.), *Marinetti: Selected writings* (pp. 56–58). New York: Farrar, Straus and Giroux. (R. W. Flint & A. A. Coppotelli, Trans.).

Marinetti, F. T. (1911a/1971). The birth of a Futurist aesthetic. In R. W. Flint, (Ed.), *Marinetti: Selected writings* (pp. 80–83). New York: Farrar, Straus and Giroux. (R. W. Flint & A. A. Coppotelli, Trans.).

Marinetti, F. T. (1911b/1971). Electrical war (A Futurist vision-hypothesis). In R. W. Flint, (Ed.), *Marinetti: Selected writings* (pp. 104–108). New York: Farrar, Straus and Giroux. (R. W. Flint & A. A. Coppotelli, Trans.).

Marinetti, F. T. (1913/1973). Destruction of syntax—[Wireless imagination]—Words-in-freedom. In U. Apollonio (Ed.), *Futurist manifestos* (pp. 95–106). Boston, MA: MFA Publications. (R. W. Flint, Trans).

Marinetti, F. T. (1914/1987). *Zang tumb tuum.* In R. J. Pioli, (Ed.), *Stung by salt and war: Creative texts of the Italian avant-gardist F.T. Marinetti.* New York, NY: Peter Lang. (R. J. Pioli, Trans.).

Moffett, C. (1899). Marconi's wireless telegraph. *McClure's Magazine, 13*(2), 4–17.

Perloff, M. (2003). *The Futurist moment: Avant-garde, avant guerre, and the language of rupture.* Chicago: University of Chicago Press. (Original work published in 1986).

Poggioli, R. (1968). *The theory of the avant-garde.* Cambridge: Belknap Press of Harvard University Press. (G. Fitzgerald, Trans.).

Rainey, L. S. (1998). *Institutions of modernism: Literary elites and public culture.* New Haven: Yale University Press.

Rye, J. (1972). *Futurism.* London: Studio Vista.

Snyder, C. (1902). America's inferior position in the scientific world. *The North American review, 174*(542), 59.

Taylor, C. J. (1974). *Futurism: Politics, painting, and performance.* Ann Arbor: UMI Research Press.

Taylor, F. W. (1967). *The principles of scientific management.* New York: Norton. (Original work published in 1911).

White, J. J. (1990). *Literary Futurism: Aspects of the first avant garde.* Oxford: Clarendon Press.

Woolf, V. (1927/1989). *To the Lighthouse Virginia.* Philadelphia: Harvest Books.

Chapter 6
Conclusion—STS and Technical Communication: Expansive Possibilities

The treatment of scientific knowledge as a social construction implies that there is nothing epistemologically special about the nature of scientific knowledge: It is merely one in a whole series of knowledge cultures (including, for instance, the knowledge systems pertaining to "primitive" tribes). Of course, the success and failures of certain knowledge cultures still need to be explained, but this is to be seen as a sociological task, not an epistemological one.

Trevor Pinch and Wiebe E. Bijker
(as cited in Pool 1997, p. 13)

Science and Technology Studies has had an impact on technical communication scholarship/pedagogy over the last three decades. I believe my analysis of Marconi's wireless is an example of a project that is essentially about scientific and technical communication but influenced heavily by cultural studies approaches of STS. No longer is technical communication simply "nuts and bolts"; instead, technical communication pedagogy approaches research from historical, philosophical, and social perspectives. This approach is best shown through Rivers' (1994) call for a more robust technical communication discipline: He argued "[technical communication scholars] need a better understanding of the interactions between technical, scientific, and business communication done in different languages and cultures" (p. 45). He also supported using approaches to technical communication scholarship that lay outside the field; in fact, he specifically mentioned the disciplines of literature and literary criticism as two areas to "incorporate into our bag of analytical tools" for studying technical communication (p. 45). Although Rivers' call is from the mid-1990s, the field of technical communication has not fully embraced cultural analysis. A brief look at the field's major journals over the past decade shows the field embracing more instrumentalist approaches to research and workplace analyses. Scholars rarely discuss rhetoric beyond rhetorical strategies aimed at workplace audiences; therefore, *rhetoric* as building meaning into concepts at a cultural level, for instance, how culture is reproduced in technical/scientific documents, is neglected by the field.

This analysis on Marconi's wireless addresses Rivers' call because the invention was a product not just of a lab but of a culture. The wireless existed because rhetoric shaped its value and viability in relation to cultural attitudes, values, and practices. The wireless was popular and successful because relevant social groups fit it into their time period's desires. However, lay audiences, believing in the

A. A. Toscano, *Marconi's Wireless and the Rhetoric of a New Technology*,
SpringerBriefs in Sociology, DOI: 10.1007/978-94-007-3977-2_6,
© The Author(s) 2012

neutrality of language, do not always notice that discourse "constructs" technologies and sciences. One popular notion of technical and scientific communication is that those discourses communicate facts with singular interpretations derived effortlessly from experiments; unfortunately, such a perspective ignores the social construction of science and technology. Just as famous scientists such as Einstein, Watson, Crick, Curie, and Salk had to communicate their findings to receptive audiences through rhetoric, Marconi and pro-Marconi writers had to employ rhetoric to construct the wireless as a viable technology that fit within industrial cultures. Popular representations aided the rhetorical construction of Marconi's wireless thus creating interest in the invention. The wireless was progress, evolution, efficiency; it had an image of prosperity and viability before it became a black box technology. As a cultural artifact, the wireless, or rather, the success of the wireless, suggested that the discourse surrounding it and the culture's overall fascination with technology helped audiences believe in the wireless and, therefore, realize it. Furthermore, Marinetti's fetishization of the wireless demonstrated a cultural fascination but through an exaggerated medium—avant-garde art. Without positive discourse, the wireless would not have become a realized technology, and, without its reinscription into Futurist aesthetics, it would not have been able to represent the early twentieth century's value for *hyper*-industrialization. After all, Marinetti's love of efficiency through art is not very different from Taylor's blueprint for industrial efficiency.

Although this book closely analyzes many texts about the wireless, further analysis of other texts would enrich this study. Patent documents or other technical specifications would offer revelations on technical writing conventions. Focusing on business communication surrounding the wireless, such as marketing texts, business proposals, and shareholder reports, could help analyze the rhetorical strategies employed in that particular discourse convention. However, the wireless was a product that was more than a "sound," profitable technical application; it was an invention that held many early twentieth-century values.

The wireless did permeate Western culture in the early twentieth century, and there are more tropes of progress associated with wireless descriptions. Being labeled as a progressive technology suggests the wireless "spoke" to the cultural desire to advance through machines. Relevant social groups promoted the wireless as an evolutionary feat, a marker of industrial progress, just as Edison's electrical works projects were seen in places like San Francisco and Louisville (Bazerman 1999, p. 219). Analyzing such cultural values is the STS influence of this book, which I hope expands research in technical communication. Therefore, the pedagogical base of technical communication scholarship should expand to incorporate the theories and methodologies of STS, especially the case study approach for analyzing the rhetorics of technologies and sciences. Technical communication students can benefit from work outside of "nuts and bolts" discussions of layout and design. If technology is a semiotic system, one that we can read, bringing broader cultural concerns into the classroom ought to enable discussions and further inquiry into technical communication history.

I believe this book's investigative approach to Marconi's wireless can be expanded or redirected in several ways to apply better to classrooms. First, more emphasis should be placed on pedagogical implications of analyzing the discourse of an historical phenomenon/apparatus. Students should learn that science and technology do not exist without rhetoric: Physical inventions and impressive theories are not realizable without community acceptance. Second, tracing the discourse surrounding radio innovations after Marconi's "black box" would allow for an important historical analysis of the trends of rhetoric surrounding the radio. Such a direction would help demonstrate the way(s) in which progress continued to be associated with this invention or even mass communication in general. Third, contrasting pro-Marconi rhetoric with anti-Marconi rhetoric and the rhetoric surrounding other wireless pursuits of the early twentieth century would help identify instances where Marconi's wireless's rhetoric was similar or dissimilar to that of contemporary inventions. This research would help readers better understand why Marconi is historically seen as *the* inventor of the radio even though he mainly assembled the apparatus. His celebrity status as being the first to excite the world with his invention is a major reason for Marconi's title of "father of the radio." Finally, each chapter could be a book in and of itself that analyzes more instances where the wireless's rhetoric and literary value demonstrate why the invention was a *monumental* early twentieth-century technology that embodied particular tropes of progress—speed, efficiency, evolution, ahistoricity, and profitability.

We can always "grab more actors" as Latour (1996) says. These new actors will tell us new stories and support former positions. Technologies become realized because potential users perceive or are made to perceive their usefulness. Society accepts technologies because they become familiar to our values and practices. Likewise, society rejects technologies when they disrupt life or do not fit into the ideology of the time and place. Even though individuals may reject certain technologies, society may collectively accept technologies perceived as "necessary." As Bazerman (1998) argued, "products of the built environment" reflect cultural values. We learn much about a culture by analyzing its technologies and the discourse surrounding those technologies.

References

Bazerman, C. (1999). *The languages of Edison's light*. Cambridge: MIT Press.

Bazerman, C. (1998). The production of technology and the production of human meaning. *Journal of Business and Technical Communication, 12*(3), 381–387.

Latour, B. (1996). *Aramis, or the love of technology*. Cambridge: Harvard University Press. (C. Porter, Trans.).

Pool, R. (1997). *Beyond engineering: How society shapes technology*. Oxford: Oxford University Press.

Rivers, W. E. (1994). Studies in the history of business and technical writing: A bibliographic essay. *Journal of Business and Technical Communication, 8*(1), 6–57.

Chapter 7
Notes

1 Introduction

1. Because this book argues that seemingly intangible forces led to the development of technologies and that individual inventors were merely responding to the demands of a time period, I use several passive voice constructions to convey agent-less action. For instance, "the rhetoric of the wireless was mediated by the time period's ideology" is a passive voice construction, but, in the context of a socially constructed artifact, the wireless and other technologies come to be because they fulfill a demand in society. Marconi and other supporters have some agency in creating a viable technology, but the technology first had to adhere to social needs. Therefore, it would be difficult and misleading to have active voice constructions for all statements about how the cultures of early twentieth century industrial nations mediated the rhetoric of the wireless.
2. Because I argue that the journalists are a mass relevant social group, I don't introduce their names in the sentences; instead, I include their names, if there is an author, in the in text citations. Rhetorically, I am asking readers to not concentrate on the journalists similarly to how they might concentrate on the scholars I cite. The reason is an attempt to read the journalists together and not individually. In most cases, I use the phrase "an article," "another article," "one article," and related phrases to refer to journalists' texts.

2 Chapter 1

1. The phrase "technical writing" has given way to the phrase "technical communication" in the last two decades. I use the phrase "technical communication" in this book, but, when quoting other sources, I use the author's exact phrase.

A. A. Toscano, *Marconi's Wireless and the Rhetoric of a New Technology*,
SpringerBriefs in Sociology, DOI: 10.1007/978-94-007-3977-2_7,
© The Author(s) 2012

3 Chapter 2

1. Johnsom (1995) is Bruno Latour's pseudonym in the article "Mixing Humans and Nonhumans Together."
2. Similarly, Bijker (1995) analyzes the social relations and historical discourse related to three technologies—bicycles, bakelites, and light bulbs—to demonstrate how those inventions were shaped by social attitudes as much as they were shaped by the inventors.
3. For an analysis on Foucault's contribution to philosophy of science, see Rouse's (1987) *Knowledge and Power*. Rouse argues from the Foucauldian perspective that group interaction influences the products created:

   The various ways people are enclosed, grouped, distributed, separated, and partitioned mark a related spatial organization of power/knowledge. These distinctions constrain our patterns of activity and interaction, and in doing so they shape both our activities and us as agents. (p. 217)

4. Using the word "discoveries" for scientific advancement may be misleading because the connotation is that the science or technology is out there waiting for someone to find it. So-called discoveries are really the process of actors working together or, at least, interacting with experiments and research. To claim "discovery" as opposed to "process" is to assume a whig history of technology, a narrative that claims the process of creating science and technology flowed seamlessly, thus, ignoring the starts and stops and reinterpretations of the technology or science by the actors. For more discussion on STS and whig histories, see Staudenmaier (1985).
5. Hiskes and Hiskes (1986) divide historical approaches of technology into "four historical periods: (1) ancient Greece through the fifteenth century, (2) the Scientific Revolution from 1540 through 1750, and (3) 1750–1940"; the "somewhat speculative" fourth period starts in 1940 and goes to the present, which was 1986 (p. 21). In each "period" science and technology vary in their relationship to each other. For instance, in ancient times science and technology were distinct but "very intimate" disciplines during the Scientific Revolution (pp. 21, 22). Hikes and Hikes do mention, however, that "[a]lthough the scientific and technological communities were isolated from each other during the 1750–1940 periods…occasional transfers of knowledge between them did occur" (p. 24).
6. This situation allowed the Internet to flourish. Also, a culture that values consumerism would use this technology to facilitate consumer purchasing.
7. However, a science such as evolution has much more to do with the public. Currently, major challenges to studying evolution in public schools show the politics of a socially constructed science. Many school districts have to contend with the possibility of indoctrinating students in "intelligent design" alongside the science of evolution. The major participants are not only scientists but

parents, politicians, and judges. Consequently, if social pressure questioning evolution gets school districts to demand such content, publishers will have to adjust their textbooks to include "scientific" discussions of creationism, currently under the guise of "intelligent design"; therefore, we see the market as a force in establishing science and communicating technical information. The wireless had to be a market-oriented technology in order to be accepted as well.

8. Ritzer (1996) argued contemporary cultural systems such as McDonald's ultra-efficient business model promote the "rationality of irrationality": "[R]ational systems are unreasonable systems that deny the humanity, the human reason, of the people who work within them or are served by them. In other words, rational systems are dehumanizing" (pp. 123, 124).

9. For example, Chinese and Indian *modernization* relates to the two countries' acquisition of industrial tools that have allowed their economies to grow.

10. This analogy works best for presidential elections and not state and local elections. In order to run successfully for president in modern times, candidates must be able to secure tens of millions of dollars in campaign funds. That process is much easier with support from either the Democratic or Republican parties. Occasionally, third party candidates can receive enough votes to hurt other candidates if their political beliefs are closely aligned (e.g., Ross Perot taking votes from George H. W. Bush in 1992), but the outcome is the still the same—either candidate A or B wins.

11. Feenberg also ignored the fact that the Internet was one of many post-Sputnik influenced technologies created to establish America's scientific, militaristic, and economic dominance through sharing scientific information with allies, bases, and universities.

12. The phrase "user friendly" is extremely important to understanding how computers shape social practice. Before graphical user interfaces (GUIs), such as Microsoft Windows, computer users had to understand the language of computers through text-based commands. To learn these commands, one had to learn a new language. Currently, this language has been converted to graphical object-oriented commands that can be done by dragging and dropping. For instance, copying the file "chapter2.doc" from the Desktop to the My Documents file folder can be done by simply dragging "chapter2.doc" from the Desktop to the My Documents folder. The text-based command would look like this:

COPY C:\Documents and Settings\Username\Desktop\ chapter2.doc
C:\Documents and Settings\Username\My Documents\

Although the above command is a relatively easy one, a user in a text-based computer environment had to enter commands similar to the one above in order to set preferences, run programs, scan for viruses, and do other "everyday" tasks. Those text-based commands are now done automatically or with the (double) click of the mouse. Each new computer "advancement," whether it be hardware or software, is marketed as more user friendly; therefore, to "fix" a computer, users need only buy the next release and not understand the complex programming behind it or previous procedures for accomplishing tasks.

13. An exception to this is the growing open source software available to users. Open source means that software is usually free or inexpensive. Operating systems such as LINUX are examples. However, users normally use open source in conjunction with proprietary software, such as Windows or Macintosh.

14. Brandt (2001), although not specifically addressing computer literacy but *literacies*, argued that literacy is a basic civil right (p. 206).

15. The fact that we could identify predominantly African-American communities (as opposed to "human" communities) attests to systemic and historic racism.

16. Although *Brown v. the Board of Education of Topeka Kansas* overruled *Plessey v. Ferguson*, institutionalized racism still exists. Democracy is often associated with freedom and egalitarianism, but larger cultural values influence how citizens vote; historically, voting in America has been to the detriment of African Americans from slavery to Jim Crow. Additionally, homophobia and heterosexism have been "sanctioned" by voters in many states because they vote to outlaw the right for same-sex couples to marry.

17. Theodore Roosevelt had been trying to build up the navy since he was the Assistant Secretary of the Navy under William McKinley; he "observ[ed] the success of Great Britain, the growth of Germany, and the increasing dominance of Japan as a Pacific power," and he believed the U.S. "must also have a great navy" (Nelson 2001, p. 3). In the historical moment of the Spanish-American War and the decade after, modernizing the navy fit with expansionist goals. From the point of view of a certain relevant social group for wireless technology being important for the navy, Roosevelt and senior officials helped shape the new navy through new technology.

18. Even though Marconi's company did not win the bid, Marconi was the first to outfit U.S. Navy ships. Marconi was "invited to install wireless telegraphy sets on two warships for onboard testing [in 1899]" (Yeang 2004, p. 2). By Marconi's (1900) own admission, he chose not to use his most up-to-date device during the U.S. Navy tests because he had not yet secured an American patent (p. 294), an obvious political and economic concern.

19. Williams (2000) did not consider financial systems or other abstract computer systems as technologies (p. 661). She appeared to express nostalgia for past "true" engineering, which according to her was "as much about creating a moral world as it was about creating a world of knowledge or of things" (p. 658). She also "mused" about whether or not "technological development can continue to flourish indefinitely in a market environment" (p. 664). She briefly discussed open source as a potential departure from capitalism and (possibly) to more democratic control because "[o]pen-source advocates believe that truly robust software can be built only when it is treated as a form of scientific knowledge, so that incremental contributions accumulate indefinitely through a system of peer review" (p. 664). However, she claimed "technology that profits from change for the sake of change" is hardly a "proper" situation (p. 664).

20. In Huxley's (1932/1960) *Brave New World*, the characters use "My Ford" instead of "My God" to show disbelief. This apotheosis of Ford is Huxley's interpretation that society has embraced industrialization as a religious or spiritual value.
21. White (1990) suggested that Marinetti's claim that his prose style was influenced by reporting his stories using Morse code while he was a war correspondent during Italy's bombing of Tripoli was a fabrication (p. 162).
22. Although humans (or, *hominids*, more specifically) have a common ancestor with modern apes, humans did not evolve *from* apes. Such a view assumes that modern apes never evolved but, instead, stayed in the same form as they are today—not undergoing natural selection or adaptation. In other words, evolution is not a *progressive* change as popular definitions would have us believe.
23. And "machine-man" is more appropriate for Marinetti than "machine-human" because of his disdain for women. Although Hewitt (1993) argued that phallocentrism is indicative of fascist ideology (p. 153), I argue industrial ideology is more relevant to Marinetti's early Futurist work than reading it as proto-fascist.

4 Chapter 3

1. This presentation is published in two parts in *The Electrical Review*. Part I is published in the 15 June 1901 issue, and part II is published in the 22 June 1901 issue. Because both are the same presentation—"Syntonic Wireless Telegraphy"— I cite them as the same source, but they are listed separately in the References.
2. In order to limit confusion in mistaking one presentation for another, I include the date of Marconi's presentations in each in-text citation even if I already note the presentation's date once in the paragraph.
3. A *coherer* is a vacuum tube filled with pieces of iron and used to detect radio waves. Marconi used Edouard Branley's coherer in his early experiments, which was a great improvement over the older coherer invented by Heinrich Hertz. This is one reason to consider calling Marconi the "assembler" of the radio because he did not invent each component. Instead, he put the components together and envisioned that wireless signals could travel to and from transmitters and receivers.
4. Here I must pause to emphasize that this is a step toward decreasing human involvement in technical systems. Automation eventually removes human inefficiency (and error) by literally removing humans from tasks, but, before total human removal occurs, operational tasks are increasingly reduced to repetitive, simple commands or movements that cause the operator to work in a mechanical fashion.

5. Marconi established this law in a presentation to the Institution of Electrical Engineers—which is now know as the Institute of Electrical and Electronic Engineers (IEEE)—in March 1899 and restated the old law in this presentation. The law read as follows:

> [T]he distance over which it is possible to signal with a given amount of energy varies approximately with the square of the height of the vertical wire, and with the square root of the capacity of a plate, drum, or other form of capacity area. (1900, p. 289)

6. "The daylight effect" is the phenomenon whereby radio signals travel farther at night than during the day. Marconi and others knew about this, but did not understand why. In his Nobel Prize lecture, Marconi believed this effect had something to do with clear skies (1909, p. 211). However, long-range radio signals travel farther and stronger due to the fact that radio waves bounce off the Earth's ionosphere better at night. This does not work, however, for later FM transmissions.

7. Technology "working itself out" simply means technologies will eventually be able to do more and more things for users. For instance, the automatic spell checker has found a way to "fix" user spelling errors. Of course, the autocorrect function is not perfect because it often "corrects" what users do not want changed. On a cultural level, however, the above assumption that computers will make life easier in the future is an ingrained positive view of technology— future technologies will make life easier. That follows the widely held belief that technology *always* improves.

5 Chapter 4

1. Mifeprestone (or RU-486) is a good (albeit controversial) example because its introduction in the United States had to overcome social, political, and scientific pressure that thwarted its approval for many years.

2. Although the US Navy appears to have awarded a contract to an American over a foreign company for certain experiments circa 1910 (c.f., Yeang 2004), Marconi's status as an inventor was quite high in the American popular press.

3. The Marconi Signal and Wireless Co, founded in the late nineteenth century, continued to have "Marconi" in its name—there were minor changes—until the company was sold off in January 2006. Rogue polemicists do rail against Marconi as the "father of the radio" and advance the idea that Nikola Tesla was the *true* inventor, and Marconi was a thief. Of course, trading one lone inventor for another ignores the larger social context under which the wireless came to be.

4. Alexander Stepanovitch Popoff, a Russian electromagnetic scientist in the late nineteenth and early twentieth century, has his name spelled "Popoff" and

"Popov" in journalistic and historical accounts. When I quote an author's mentioning of him, I use the same spelling as the author.

5. A current example of how a specific name secures itself as a benchmark long after other companies' products have reached the market is the adjective/verb/noun XEROX: Many of us have created *xerox* copies, *xeroxed* originals, or asked "Where's the *Xerox*?" in order to photocopy originals using an HP, Panasonic, Samsung, or other photocopier brands.

6. Arthur C. Clarke (2000), the famous science fiction writer, claimed that "Any sufficiently advanced technology is indistinguishable from magic" (p. 2).

7. J. Chandra Bose has recently been accepted as the inventor of the coherer Marconi used in his 1901 transatlantic experiments (Bondyopadhyay 1998, p. 259).

8. Jabbari (1997), a senior IEEE member, noteed that "Marconi put to work the concepts already developed by Hertz, Branly, Lodge, Popov, and a few others" (p. 1523).

9. Hancock's (1974) history of Marconi's wireless, written in the late 1940s as a celebration of Marconi's work, also overlooked Marconi's contemporary contributors.

10. "Ether" was believed to be the unseen matter in the universe, a substance that could conduct electricity and, subsequently, radio waves.

11. While it may seem as if I am glorifying Marconi and contributing to his lone inventor legend, my emphasis on him reflects my argument that the popular press rhetorically constructed the Marconi wireless apparatus as *the* most important and dominant system. His celebrity status and (eventual) Nobel Prize—which he shared with a lesser-known German inventor, Karl Ferdinand Braun—does not prove historically that he was the sole inventor of the wireless technology, but his popularity does suggest scientists and lay audiences alike considered his wireless to be one of the first and most important inventions.

12. Baker's (1902) introduction to this article also appeared in the March 1902 issue of *Current Literature*, and another large section is quoted in Herbert Wallace's June 1902 *McClure's Magazine* article.

13. *McClure's Magazine* was extremely positive about Marconi's work. Robert McClure was actually present in Dover, England for Marconi's crossing of the English Channel and was "allowed to hold cross-channel conversation" with Cleveland Moffett in order to, "in the interests of [*McClure's*] readers, satisfy ourselves that this wireless telegraphy marvel had really been accomplished" (Moffett 1899, p. 15).

14. Although the Internet in the form of the World Wide Web was recognized by consumers in the early 1990s, its creation dates back to the 1960s. Beaming nightly news into the homes of families seems more an invention for satellite television, but Baker (1902) believed the news would be written down for the occupants to read at their leisure. Interestingly, the idea of people gathering around a radio set or television set to listen or watch "news" was not part of the popular consciousness as evidenced in these popular press articles.

15. The popular press varied in its price-per-word estimates. Authors claimed the system would be profitable at less than a cent a word to around five cents a word. Of course, all claimed the wireless rate was significantly cheaper to cable telegram rates. Also, there were differences in the exact price of building stations and outfitting ships.

6 Chapter 5

1. Huyssen (1986) noted that "[m]odernists such as T. S. Eliot and Ortega y Gasset emphasized time and again that it was their mission to salvage the purity of high art from the encroachments of urbanization, massification, technological modernization, in short, of modern mass culture"; however, the avant-garde wished to overthrow such high-minded pretension related to "forms of bourgeois society" (p. 163).
2. Robert Frost's (1916) poem "The Line-Gang" responds to the encroachment of telephone wires from the city on the New Hampshire countryside (Rhodes 1999, p. 62).
3. Even Lawrence (1914/1979) was familiar with the Futurists and claimed he had an interest in their works: "I have been interested in the futurists....I read Marinetti's and Paolo Buzzi's manifestations and essays and Soffici's essays on cubism and futurism. It interests me very much" ([Letter Arthur McLeod], p. 180).
4. I refer to the beginning phase—or, as Bondenella and Bondenella (1979) called it, "the truly revolutionary phase" (p. 229)—as the work done from 1909 to the beginning of WWI. Marinetti's later work experiments more with performance than manifestos. Also, the movement was co-opted by the fascists following WWI and, therefore, ceased to be "edgy" or avant-garde because the movement was now part of the establishment.
5. The concept *parole in libertá* is commonly translated as "words in freedom" or "words in liberty." Either is acceptable, but readers should not confuse *parole in libertá* with "free verse"—a composition Marinetti despised as reminiscent of the past. Marinetti's goal was to free words from their syntactic "oppression" and let them communicate their essence. According to Butler (1994), "Marinetti's aim, through the disruption of syntax, metre, and punctuation, is to produce a 'lyrical intoxication', which will abolish the reassuring musical continuities of *vers libre*, in favour of an abrupt, instantaneous, telegraphic form of communication" (p. 173).
6. Although Marinetti was the first Futurist and Futurism was the first avant-garde movement, other futurisms flourished in Europe in the early twentieth century—Russian Futurism, Ego-Futurism, and Cubo-Futurism. Even Vorticism, an English avant-garde led by Wyndam Lewis, resembled Italian

Futurist aesthetics. Futurism's appeal could be found in places like Italy, France, Poland, Germany, and Russia (Calinescu 1987; Childs 2000; Nicholls 1995; Perloff 2003; White 1990). But, as Perloff (2003) argues, the avant-garde's short-lived internationalism ended with the onset of nationalist fervor leading up to WWI (p. xxxvi).

7. Marinetti's incendiary work was a major component of his public perfor-mances, such as his "attack" on Venice (Perloff 2003, pp. 201–202) and the packed theater houses that stunned Ezra Pound (Rainey 1998, p. 29). This violence and discordant noise—as the Futurists often brought loud speakers in order to play awful sounds for the audience—continued into the 1920s with the Variety Theater.

8. "Radio-bomb" is a phrase I use to demonstrate Marinetti's desire to assault audiences with his manifestos and his love of destruction. His texts reveal a strange desire to be beamed into the physical minds of audiences. The ultimate "evolution" for him would be to actually become a radio wave to be broad-casted to the world.

9. Technology was not the only movement organized around nationalism. A political rally for Italy's intervention into WWI actually started the acquaintanceship between Marinetti and Benito Mussolini. Their relationship began around 1915 when the two were arrested for an interventionist rally (Bondenella and Bondenella 1979, p. 318). The two men banded together possibly because of Marinetti's anti-communism and his acceptance of Mussolini's nationalist sentiments for a strong, united Italy. Today, the fascist party in Italy (led by Mussolini's grand-daughter) still believes in a strong free market expansionism for the modern Italian state, but they differ from their close right-wing allies in parliament, *La Lega Nord* (LN), on one very fun-damental issue—Italian unification. The LN wants to divide Italy into separate northern and southern nations because the industrialized north feels the "backward" south is a blight to economic prosperity. The fascists, on the other hand, do not want a divided Italy.

10. Sociologist Ritzer's (1996) book *The McDonaldization of America* describes how these rational ultra-efficient practices have crept into contemporary ser-vice-economy jobs.

11. Translated as POETRY BEING BORN (Marinetti 1914/ 1987 p. 57).

12. Carlo Carrá left the Futurists in 1916 after WWI (Rye 1972, p. 153). Although Giovanni Papini was a lesser-known Futurist—who joined the movement with Soffici in 1913 (Rye, p. 13)—the eighth soul could have very well been Benito Mussolini's because the novel was written before Marinetti's disillusionment with the fascist dictator. Also, the soul could have been Armando Mazza, another lesser known Futurist poet.

13. I call this work a "poem-manifesto" because it has polemic manifesto qual-ities much like "The Founding and Manifesto of Futurism" (1909/1917), and

it also has poetic wireless imagery—energy, kilowatts, dials. However, unlike other manifestos, Marinetti did not set down specific goals for his movement, and, unlike his heavily onomatopoeic poems, this text described energy, bombardment, and movement using common descriptive adjectives and not made-up words or extravagant typefaces and font sizes.

14. This quotation encapsulates White's (1990) argument that the Futurists glorified the science fiction of electrical forces of the imagined twenty-first century.

15. By Objectivist, I mean Ayn Rand's ultra-capitalist philosophy advocating "the virtue of selfishness."

16. Another science fiction example of "reading" or perceiving electrical forces occurs in *The Matrix* (1999). To observers outside the Matrix, the binary code flashing on monitors represents action inside the Matrix, which is a "fake" world created by machines.

17. Hewitt (1993) covered Marinetti's and other modernists' fascistic ideologies in his book *Fascist Modernism*.

18. Interestingly, Snyder's (1902) comment follows a list of prominent Europeans who "made wireless telegraphy possible" (p. 59). Also, Snyder points to "Marconi's admirable triumphs" as an extension of past scientists' work in electromagnetism (p. 59).

19. Apparently, "the rapport of centuries between poet and audience" is an element or relationship of the past that Marinetti wishes to keep (1913/1973, p. 98).

20. Seven of Nine, a popular cyborg from *Star Trek Voyager* (Berman, Taylor, Biller, and Braga 1995–2001), was rescued from the efficient Borg collective. And John Flynn from *Tron* (Kushner 1982) is digitized in order to enter the computer; thus, in order to be replicated, humanity must become machine-like before reproduction.

References

Baker, R. S. (1902). Marconi's achievement: Telegraphing across the ocean without wires. *McClure's Magazine, 18*(4), 4–12.

Berman, R., Taylor, J., Biller, K., Braga, B. (1995–2001). Star Trek Voyager [Television Series]. United States: Paramount Television. (Producers).

Bijker, W. E. (1995). *Of bicycles, bakelites, and bulbs: Toward a theory of sociotechnical change*. Cambridge: MIT Press.

Bondenella, P., & Bondenella, J.C. (1979). *Dictionary of Italian literature*. Westport: Greenwood Press.

Bondyopadhyay, P. K. (1998). Sir J. C. Bose's diode detector received Marconi's first transatlantic wireless signal of December 1901 (The 'Italian navy coherer' scandal revisited). *Proceedings of the IEEE, 86*(1), 259–285.

Brandt, D. (2001). *Literacy in American lives*. Cambridge: Cambridge University Press.

Butler, C. (1994). *Early modernism: Literature, music and painting in Europe, 1900–1916*. New York: Clarendon Press.

Calinescu, M. (1987). *Five faces of modernity* (2nd ed.). Durham: Duke University Press.

Childs, P. (2000). *Modernism*. London: Routledge.

Clarke, A. C. (2000). *Profiles of the future*. London: Indigo.

Hancock, H. E. (1974). *Wireless at sea*. New York: Arno. (Original work published 1950).

Hewitt, A. (1993). *Fascist modernism: Aesthetics, politics, and the avant-garde*. Stanford: Stanford University Press.

Hiskes, A. L., & Hiskes, R. P. (1986). *Science, technology, and policy decisions*. Boulder: Westview Press.

Huxley, A. (1960). *Brave new World*. New York: Harper. (Original work published 1932).

Huyssen, A. (1986). *After the great divide: Modernism, mass culture, postmodernism*. Bloomington: Indiana University Press.

Jabbari, B. (1997, October). Introduction to the classic paper by Marconi. *Proceedings of the IEEE, 85*(10), 1523–1525.

Johnsom, J. (1995). Mixing humans and nonhumans together: The sociology of a door-closer. In S. L. Star (Ed.), *Ecologies of knowledge: Work and politics in science and technology* (pp. 257–277). Albany: State University of New York Press. [a.k.a. Bruno Latour].

Kushner, D., & Lisberger, S. (1982). *Tron [Motion picture]*. United States: Buena Vista Distribution. (Producer).

Lawrence, D.H. (1914/1979, June 2). [Letter to Arthur McLeod]. Letter 731. In G.J. Zytaruk & J.T. Boulton (Eds.), *The Letters of D. H. Lawrence* (Vol. 2, pp. 180–182). Cambridge: Cambridge University Press.

Marconi, G. (1900). *Wireless telegraphy*. Smithsonian annual report, 1901, 287–296. (Original published in Proceedings of the Royal Institution of Great Britain, 16(2), 247–256.)

Marinetti, F.T. (1909/1971). The founding and manifesto of Futurism. In R.W. Flint (Ed.), *Marinetti: Selected writings* (pp. 39–44). (trans: Flint, R.W. & Coppotelli, A.A.). New York: Farrar, Straus and Giroux.

Marinetti, F.T. (1913/1973). Destruction of syntax—[Wireless imagination]—Words-in-freedom. In U. Apollonio (Ed.), *Futurist manifestos* (pp. 95–106). (trans: Flint, R.W.). Boston: MFA Publications.

Marinetti, F. T. (1914/1987). Zang tumb tuum. In R.J. Pioli (Ed.), *Stung by salt and war: Creative texts of the Italian avant-gardist F.T. Marinetti*. (trans: Pioli R.J.). New York: Peter Lang.

Moffett, C. (1899). Marconi's wireless telegraph. *McClure's Magazine, 13*(2), 4–17.

Nelson, K. (2001) Theodore Roosevelt the navy and the war with Spain. In E.J. Marolda (Ed.), *Theodore Roosevelt the U.S. Navy and the Spanish-American War*. New York: Palgrave (pp. 1–6).

Nicholls, P. (1995). *Modernisms: A literary guide*. Berkeley: University of California Press.

Perloff, M. (2003). *The Futurist moment: Avant-garde, avant guerre, and the language of rupture*. Chicago: University of Chicago Press. (Original work published in 1986).

Rainey, L. S. (1998). *Institutions of modernism: Literary elites and public culture*. New Haven: Yale University Press.

Rhodes, R. (Ed.). (1999). *Visions of technology: A century of vital debate about machines, systems and the human world*. New York: Touchstone.

Ritzer, G. (1996). *The McDonaldization of society*. Thousand Oaks: Pine Forge Press.

Rouse, J. (1987). *Knowledge and power: Toward a political philosophy of science*. Ithaca: Cornell University Press.

Rye, J. (1972). *Futurism*. London: Studio Vista.

Snyder, C. (1902). America's inferior position in the scientific world. *The North American Review, 174*(542), 59.

Staudenmaier, J. M. (1985). What SHOT hath wrought and what SHOT hath not: Reflections on twenty-five years of the history of technology. *Technology and Culture, 25*(4), 707–730.

White, J. J. (1990). *Literary Futurism: Aspects of the first avant garde*. Oxford: Clarendon Press.

Williams, R. (2000). All that is solid melts into air: Historians of technology in the information revolution. *Technology and Culture, 41*(4), 641–668.

Yeang, C.-P. (2004). Scientific fact or engineering specification? The U.S. Navy's experiments on wireless telegraphy circa 1910. *Technology and Culture, 45*(1), 1–29.